ECOLOGICAL RESTORATION

ECOLOGICAL RESTORATION

GEORGE H. PARDUE

AND

THOMAS K. OLVERA

EDITORS

Nova Science Publishers, Inc.
New York

NOTICE TO THE READER

LIBRARY OF CONGRESS CATALOGING-IN-PUBLICATION DATA

Pardue, George H.
  Ecological restoration / George H. Pardue and Thomas K. Olvera.
     p. cm.
  Includes index.
  ISBN 978-1-60741-013-3 (hardcover)
  1. Restoration ecology. I. Olvera, Thomas K. II. Title.
  QH541.15.R45P365 2009
  333.71'53--dc22
                                                                        2009000176

Published by Nova Science Publishers, Inc. ✦ New York

CONTENTS

PREFACE

Human exploitation of stream and land ecosystems has created the need to preserve, restore and rehabilitate. This book discusses the scientific basis of restoration for ecological improvement and some key issues associated with overall projects. In addition, some of the challenges and opportunities for further research in river restoration science is explored and the associated fields of ecohydraulics and eco-hydromorphology. An analysis of the effects of river side-arm restoration on ecosystem functions is also included, as well as an experimental approach to assessing marsh restoration as compared to a strict monitoring process. The concept of combinatorial biodiversity auction is discussed in this book, showing that the combinatorial biodiversity auction can allow farmers to benefit from cost complementarities and help to maximize biodiversity outcomes within a given budget. Also included is a presentation of the need for restoration and rehabilitation projects developed according to an ecological theory considering species life history, habitat template and spatio-temporal scope, the bi-directional effect of self-renaturalisation processes on the environment and a study on how different river typologies may alter the evaluations of biological quality.

Chapter 1 - Conservation of native biodiversity on private lands is one of the main targets of present-day environmental policies. In 1986, under its Conservation Reserve Program (CRP), the US Department of Agriculture (USDA) pioneered the use of bidding mechanism, or auctions to pay farmers to provide environmental services. The auctions offered significant advantages in terms of efficiency and cost minimization by revealing private information on opportunity costs and biodiversity resources held by landholders. This allowed the USDA to identify the most efficient projects. Auctions are now used in the European Union and Australia as part of conservation programs.

Almost all the conservation auctions implemented to date have been input-based, where farmers are paid for undertaking prescribed actions rather than committing to deliver certain biodiversity outcomes. There are high uncertainties or risks in the production of environmental goods due to the long term and stochastic nature of ecological processes. As a result, farmers might find it difficult to commit to outcomes. Also, there is a lack of understanding of the environmental impact of conservation activities, which poses difficulty in linking observed improvement in the environmental conditions with the farmer's actions. For these reasons, current auctions involve arrangements where governments pay farmers for their efforts.

However, input-based auctions may not be suitable for conservation of endangered species as an economically rational farmer will try to minimize their efforts or costs in meeting contract obligations and may not pay much attention to the agency's desired conservation outcomes. Also, the auctions provide little incentive and less scope for innovation by the farmers; farmers are limited to the intervention types specified in the projects and cannot be creative or adaptive in how they deliver biodiversity outcomes. These limitations can be addressed through output-based auctions that judge a farmer's performance by outcomes rather than specific management actions, e.g., the number of individual birds of a particular species on a property. Hence farmers are paid for ecological services. This auction type has been trialled in Germany and Australia.

However, for outcome based auctions to be successful, some design issues need to be studied. A key research gap is a lack of understanding on the degree of outcome bundling required by an agency when targeting a set of species over a landscape. This is particularly important to the farmer when the cost of conserving the species is complementary or sub-additive; i.e. the cost of delivering multiple outcomes is less than the total cost of delivering individual outcomes. This chapter focuses on this issue.

Chapter 1 - This chapter reviews conservation auction designs to assess the degree to which cost-effectiveness can be improved by allowing farmers to tender for a combination, or a package, of target species. The concept of the combinatorial biodiversity auction, in which the farmers can submit bids for any number of target species.is introduced. To demonstrate this type of auction, a bio-economic conservation model is introduced for three native endangered species (red-tailed phascogale, carpet python and malleefowl) found in the wheat-belt of Western Australia. The combinatorial biodiversity auction can allow farmers to benefit from cost complementarities and hence help the procuring agency to maximize biodiversity outcomes within a given budget.

Chapter 2 - Recent decades have witnessed a growing appreciation among the academic and practitioner communities and the wider public of the importance of the ecological goods and services provided by freshwater systems, and the extent to which they have been degraded by human activity. In the case of river channels and their floodplains, a vast and diverse range of river restoration and rehabilitation schemes have been undertaken around the world in an attempt to remedy the misuse of water resources by humans in modern times. Many invaluable lessons have been learned through the implementation of such projects, in terms of both the practical realization of restoration efforts and the science which underpins them, which have broadened and deepened the understanding of the complexity and dynamism of river systems across a wide range of spatio-temporal scales. While there is cause for much optimisim, a range of scientific, technical, social, cultural and economic barriers persist and must be fully understood and addressed in order improve restoration success rates. This chapter explores the development of river restoration through the rise of environmentalism, discusses the scientific basis of restoration for ecological improvement, and examines some key issues associated with overall project success in terms of a range of ecological and socio-economic criteria. The final sections explore some of the key challenges and opportunities for further research in river restoration science and the associated fields of ecohydraulics and eco-hydromorphology.

Chapter 3 - The deterioration of the functioning of river ecosystems has prompted rehabilitation and restoration measures during the last 20 years, based on the assessment of the ecological integrity. Most of these efforts have increased the spatial heterogeneity of these ecosystems. Yet, a more integrated approach including restoration of landscape dynamics and key ecosystem processes such as carbon and nutrient cycling is necessary. Large-scale rehabilitation and restoration projects, therefore, also consider altered nutrient dynamics or aim at reducing nutrient transport in river corridors by increasing nutrient retention. The present chapter analyzes the effects of river side-arm restoration on ecosystem functions within the side-arm, and highlight potential effects on the main channel and downstream areas in a large river. The principles of hydromorphological dynamics control nutrient cycling, phytoplankton production and interactions within the planktonic food web are demonstrated. These findings confirm the environmental control on these biological processes and their potential use as proxies to assess the consequences of hydrological changes restoration measures on river ecosystem functioning.

Specifically, the following questions are addressed related to ecological restoration of side arms: (i) what are the effects of varying hydrological connectivity on nutrient concentrations and the retention capacity?; (ii) can easily derived hydrological parameters provide meaningful information on algal development and planktonic food web interactions?; and (iii) what implications do these predictions have for future rehabilitation work and integrating these effects on larger scales (sub-catchment)?

Chapter 4 - The restoration or creation of estuarine wetlands is predicated on two presumptions: marsh habitats have significant value that if lost will produce negative effects on coastal ecosystems, and engineered wetlands are equivalent to historical marsh systems. Few studies directly have documented wetland value, but a greater number have attempted to assess the equivalency of restored and existing marshes. A majority of the restoration studies focus on monitoring the convergence of certain metrics (e.g., plant density, resident species richness) between marshes. Monitoring structural changes over time can be labor intensive, often requires decades of data before observing convergence, and potentially are confounded by ecosystem resilience, modified constraints to community change, and the existence of alternative stable states for systems. Recent advocacy of an experimental approach to assessing marsh restoration progress may offer a number of advantages over strict monitoring. An experimental approach was applied to the study of early progress in a Southeastern US created salt marsh. A 6 ha pine upland along the North River, GA was graded to intertidal elevations and planted with marsh vegetation (primarily *Spartina alterniflora*) as part of a federally mandated mitigation project. A series of experiments were conducted to assess the functional similarity of the created and an existing reference marsh along the same tidal creek. Structural characteristics of the marshes were both similar (e.g., stem densities, particulate settlement) and different (e.g., plant nearest neighbor distances, sediment organic content), but would suggest that after 3-4 yrs. the restored marsh had not progressed to the point of resembling the existing marsh. However, experimental results suggested restored and existing marshes functionally are similar. The growth and survival of caged bivalves and marked snails typically resident in marshes either were not significantly different between marshes or were greater within the restored marsh. Foraging crabs appeared to exhibit a preference for natural marsh sediments, but feeding experiments indicated similar effects on sediments from both marshes. The survival of enclosed or tethered mollusks was similar between marshes except for a single instance and suggested marshes provide the same refuge function. Application of an experimental approach led to the conclusion that the restored marsh had progressed in a relatively short time to exhibit many ecological functions existent in historical marshes. An experimental approach also should provide greater insight into the potential reasons for why created marshes may not progress to historical targets.

Chapter 5 - Anthropogenic alteration of fluvial systems has occurred extensively throughout the world. Many of these alterations, including levees, dams, and channelization, disrupt functional processes of these systems and their associated floodplains. Most prominent among these disruptions is the connectivity of the river channel to the floodplain that is critical for functions of productivity, decomposition, nutrient cycling, and the biota adapted to seasonal wet/dry periods. In addition, the effects of activities such as channelization can be exacerbated by regional geology and surrounding land-use activities, resulting in dramatic geomorphic readjustments. Currently, there is considerable interest in restoring fluvial systems and their floodplains because of their valuable functions to both ecosystems and society, including water quality enhancement, flood control, erosion control, timber production, recreational opportunities, and wildlife habitat. This chapter focuses on fluvial and floodplain processes, the effects of anthropogenic alterations these functional processes, and discusses restoration guidelines for these systems.

Chapter 6 - Humans have been exploiting stream ecosystems for millenniums. During the last two centuries, the demographic explosion resulted in the overexploitation of natural resources and channel alterations in rivers and streams from worldwide. A large number of human threats lead to ecosystem depletion. It was not until the end of the 20th century that population started to become aware of the negative consequences of their development, and thus, arose the first attempts to restore fluvial environments. Restoration in stream ecosystems entails the return to former conditions, including water chemistry, hydrologic regime, geomorphology, species composition and riparian vegetation. However, in some cases, stream ecosystems cannot be restored because it becomes too expensive and difficult or simply cannot be performed. In these cases, several rehabilitation measures could be adopted to enhance some stream ecosystem services (e.g., self-purification) and meet many stakeholder goals and objectives. Both terms, restoration and rehabilitation, can be easily confused by inexperienced water managers with the finding of esthetic values, otherwise called gardening. These erroneous restoration measures are usually restricted to merely planting trees in the riparian zone but can also involve major channel modifications. Consequently, the resulting ecosystem will need continuous intervention and the benefits of ecosystem services are neglected. To overcome this misunderstanding, restoration and rehabilitation projects have to be developed according to an appropriate knowledge of the ecological theory considering species life history, habitat template and spatio-temporal scope. All these characteristics are strongly influenced by the climate setting, which takes special relevance in the Mediterranean region because of the marked seasonality and water scarcity.

Chapter 7 - Landscape changes turned drastic in 20th Century. In the agricultural, it was challenged by the need of rapid growth of agricultural production that in turn required too wet soils to be drained. Thus land reclamation significantly changed the structure of the former landscapes. The mosaic and patchiness, comprising of natural grasslands, bushes and groves, wetlands and water-bodies, as well as numerous small land-tenures, disappeared; heterogeneity was lost. The canalization of streams, which deprived them of diversity of biotic and abiotic conditions, was followed by physical destruction of riparian zones. Erosion and deflation increased; the competence of the landscape to preserve the genetic heritage of biodiversity and prevent fluvial systems from pollution as well as nutrients to be transferred into other geo-systems was restricted. Referring to the data collected over last 15 years, the attempt was made to highlight the landscape efforts to recover. It was found that these efforts involve various components of geodynamic, eolian, hydrological, biological processes leading to soil conservation, bed process stabilization, biota diversity, sedimentation prevention and water quality improvement. These alterations are also discussed as the bi-directional effect of self-renaturalisation processes in landscape on the environment and practice, and as nature's invitation to people to collaborate improving ecological situation in agro-landscapes.

Chapter 8 - In Italy, rivers and streams show very different typologies. An analysis of lotic ecosystems in Central and Southern Italy show up how different river typologies may alter the evaluations of biological quality. Eight rivers with particular features and located in different Italian regions were monitored. Three European biological methods, all using the macro-invertebrate communities as biological indicators of water quality, have been compared from these rivers data sets. In particular, the *Indice Biotico Esteso* (IBE; Italy), Belgian Biotic Index (BBI; Belgium), and Biological Monitoring Working Party (BMW.P; Spain) were compared. Accomplished were, qualitative and quantitative samples in order to calculate index values. IBE and BBI use both the indicator taxa and the total sum of taxa in the community as common basis to reckon the index value, while BMWP scores a value for all sampled groups, determined at the family level. All these three methods transform the index values into Quality Classes, whose ranking is not so different from the Index Value. On spatial scale, comparisons were performed (where possible) within each river and among different rivers. Results seem to suggest that IBE is a quite well fitted index, and its authors are now working to improve its fitting to the peculiar conditions of Southern Italy streams. BBI instead gives excessive values due to its fitting to different river typologies. BWMP shows a trend as to increase the obtained values but it gives an acceptable value to the whole community to evaluate the biological quality of waters.

°

In: Ecological Restoration
Editors: George H. Pardue and Thomas K. Olvera
ISBN 978-1-60741-013-3
© 2009 Nova Science Publishers, Inc.

Chapter 1

AUCTION DESIGNS FOR NATIVE BIODIVERSITY CONSERVATION ON PRIVATE LANDS[1]

M.S. Iftekhar, A. Hailu and R.K. Lindner

School of Agricultural and Resource Economics,
University of Western Australia, Australia

ABSTRACT

Conservation of native biodiversity on private lands is one of the main targets of present-day environmental policies. In 1986, under its Conservation Reserve Program (CRP), the US Department of Agriculture (USDA) pioneered the use of bidding mechanism, or auctions to pay farmers to provide environmental services. The auctions offered significant advantages in terms of efficiency and cost minimization by revealing private information on opportunity costs and biodiversity resources held by landholders. This allowed the USDA to identify the most efficient projects. Auctions are now used in the European Union and Australia as part of conservation programs.

Almost all the conservation auctions implemented to date have been input-based, where farmers are paid for undertaking prescribed actions rather than committing to deliver certain biodiversity outcomes. There are high uncertainties or risks in the production of environmental goods due to the long term and stochastic nature of ecological processes. As a result, farmers might find it difficult to commit to outcomes. Also, there is a lack of understanding of the environmental impact of conservation activities, which poses difficulty in linking observed improvement in the environmental conditions with the farmer's actions. For these reasons, current auctions involve arrangements where governments pay farmers for their efforts.

However, input-based auctions may not be suitable for conservation of endangered species as an economically rational farmer will try to minimize their efforts or costs in meeting contract obligations and may not pay much attention to the agency's desired conservation outcomes. Also, the auctions provide little incentive and less scope for innovation by the farmers; farmers are limited to the intervention types specified in the

---

[1] Part of on-going PhD research of MSI. Correspondence: M. S. Iftekhar, School of Agricultural and Resource Economics, University of Western Australia, 35 Stirling Highway, Crawley, Perth, WA 6009, Australia, E-mail - mdsayediftekhar@yahoo.com

projects and cannot be creative or adaptive in how they deliver biodiversity outcomes. These limitations can be addressed through output-based auctions that judge a farmer's performance by outcomes rather than specific management actions, e.g., the number of individual birds of a particular species on a property. Hence farmers are paid for ecological services. This auction type has been trialled in Germany and Australia.

However, for outcome based auctions to be successful, some design issues need to be studied. A key research gap is a lack of understanding on the degree of outcome bundling required by an agency when targeting a set of species over a landscape. This is particularly important to the farmer when the cost of conserving the species is complementary or sub-additive; i.e. the cost of delivering multiple outcomes is less than the total cost of delivering individual outcomes. This chapter focuses on this issue.

In this chapter, we review conservation auction designs to assess the degree to which cost-effectiveness can be improved by allowing farmers to tender for a combination, or a package, of target species. We introduce the concept of the combinatorial biodiversity auction, in which the farmers can submit bids for any number of target species. To demonstrate this type of auction, we have developed a bio-economic conservation model for three native endangered species (red-tailed phascogale, carpet python and malleefowl) found in the wheat-belt of Western Australia. We show that the combinatorial biodiversity auction can allow farmers to benefit from cost complementarities and hence help the procuring agency to maximize biodiversity outcomes within a given budget.

1. INTRODUCTION

Auctions are a market mechanism with an explicit set of rules that determines resource allocation and prices on the basis of bids from the market participants (McAfee and McMillan, 1987). Bids specify payment and, in some cases, quality or quantity of the traded goods or services (Chan et al., 2003), especially where no other well-established market exists. They are used because the buyer is unsure of the opportunity cost of the bidders (Krishna, 2002). Applications of procurement auctions to biodiversity service purchases are relatively new but are attracting increasing interest from researchers and agencies (Hailu and Schilizzi, 2004).

Since 1986, under the Conservation Reserve Program (CRP), the US Department of Agriculture (USDA) has been selecting eligible farmers for land retirement contracts using a competitive bidding or auction mechanism (Latacz-Lohmann and van der Hamsvoort, 1997). In Australia and Europe, there has been an increasing interest in the use of auctions for conservation services. Auctions are important for conservation contracting for at least two reasons. Firstly, the provision of environmental services is a public-type non-market good which has no standard value. Secondly, there is two-way information asymmetry. The farmers know the costs associated with conservation measures and their impact on production and profits, whereas the government often has a better knowledge of the ecological benefits associated with the intervention (Rousseau and Moons, 2008). The bidding process can bring two parties together and reveal previously hidden or private information in a cost effective manner allowing for the identification of the most suitable projects (Latacz-Lohmann and van der Hamsvoort, 1997).

Almost all the conservation auctions implemented to date target multiple environmental benefits, ranging from simple land retirement to active wildlife conservation, and pay farmers for undertaking prescribed actions. Such umbrella input-based incentive schemes, which include input-based contracts allocated through auctions, are easy to design. Generally it is easier for the government agency to monitor project interventions than changes in environmental condition. Further, in cases where there is difficulty in linking observed changes in the environmental conditions (such as changes in the pollutant load in the rivers) with the interventions undertaken by farmers, input-based payment is often the only feasible option (Whitten et al., 2004). Even for circumstances where the procuring agency has a clear understanding of the links between conservation efforts and outcomes, the farmers might not have the same knowledge about these links and would not be able to commit to contracts specified in terms of outcomes or biodiversity benefits.

However, such input-based auctions may not be suitable for conservation of endangered species. Often endangered species are confined to specific geographic area or habitat, have peculiar traits, and are vulnerable to unique threats. Umbrella programs may not be suitable for their conservation unless interventions are particularly designed for specific species. In Europe, it has been observed that often agri-environment schemes primarily benefit common species rather than endangered and uncommon species of farmland (Kleijn et al., 2006). More successful schemes have been tailored to the needs of particular species.

To overcome these limitations, output-based auctions have been tested in Germany and Australia. Output-based auctions allocate contracts that are specified in terms of environmental or biodiversity benefits. There is a potential to improve the cost-effectiveness of conservation auctions through adopting output-based payments. For example, a Dutch non-auction based environmental conservation program in which the farmers were paid for clutches of meadow birds bred on their land found that the per-clutch payment was less expensive than compensation for agriculture production losses: payment for clutches cost 40 Euro per clutch, whereas compensating for production loss cost 100–400 Euros per clutch (Musters et al., 2001).

A key gap in the conservation auction literature is the discussion of suitable auction designs, which governments can employ to conserve a set of target species over a landscape, when the farmer's cost for conserving them are complementary or sub-additive. There are several candidate auction designs, including: sequential, simultaneous and combinatorial designs. In a sequential auction, separate auctions are run for individual goods (Grimm, 2007). In simultaneous auction, parallel auctions are run and the bidders can submit bids for several goods at the same time. Compared to the sequential auctions that buy goods one after another, a simultaneous auction permits flexible assembly as in any given round the bidder is able to focus on a set of items and bid accordingly. However, this form of bidding suffers from the 'exposure problem'. An exposure problem occurs when a bidder offers to deliver outcomes that are less costly to produce jointly, but fails to win contracts for the full set of complementary outcomes and is left with the obligation to deliver some of these outcomes without the benefit of the synergy (Xia et al., 2004). This problem can be overcome by allowing combinational bidding (Chan et al., 2003).

In a combinatorial auction, bidders can bid for any combination of goods and this allows them to better express their private information about preferences for different outcomes (Parkes, 2006). This type of auctions are frequently used in trading a variety of related goods such as blocks of radio spectrum, cars, take-off and landing slots at airports, office stationery, fishing permits, bus routes and electricity and gas transmission (Jehiel and Moldovanu, 2003). The efficiency of combinatorial auctions have been compared with other types of auctions, and it was found that over a very wide range of complementarities, combinatorial auctions perform better than simultaneous auctions, which in turn perform better than sequential auctions (Isaac and James, 2000). Combinatorial auctions for the management of additive natural resources such as allocation of areas of native timber or the allocation of aquaculture sites have been proposed (Stoneham et al., 2005). However, they have not yet been tested in conservation auctions.

In this paper, we examine the use of combinatorial auctions to conserve endangered species. In the following two sections, we review the existing conservation auction designs, and discuss the limitations of the existing designs in relation to conservation of endangered species. Then we discuss complementarity issues in conservation of multiple species. With the help of a bio-economic conservation model of three native endangered species (red-tailed phascogale, carpet python and malleefowl) found in the wheat-belt of Western Australia we demonstrate the degree to which cost-effectiveness and efficiency can be improved by allowing farmers to tender for a combination or a package of conservation outcomes. Finally, we discuss the potential use of combinatorial auction design in conservation of multiple endangered species.

2. AN OVERVIEW OF CONSERVATION AUCTIONS

Following McAfee and McMillan (1987) a conservation auction can be defined as a tender mechanism with an explicit set of rules that determines contract allocation for undertaking environmental conservation activities based on project proposals and prices from the participating farmers. Conservation auctions are used to identify farmers who can provide on-farm conservation services at minimum cost. Under such programs, the agency and the farmers have to go through a set of steps. Rolfe et al. (2008) have identified three key pre-tender activities for the agency. Firstly, the agency has to frame the auction within the current institutional and policy setting, with emphasis on identification of the level of available budget, geographic scope of the project and eligibility criteria for the participants. Secondly, an appropriate auction design has to be developed so that the farmers have proper incentives to submit feasible and cost-effective proposals. Thirdly, an appropriate selection process (metric design) should be adopted to identify the best value proposals. For a simple conservation auction, where the agency would pay the farmers for fencing remnant vegetation, the auction process would typically include the following steps (Eigenraam et al., 2006):

1. Expressions of interest— farmers located in project areas register an expression of interest.
2. Site assessments—the project officer arranges a site visit with each registered farmer. The officer assesses the site and advises the farmer on the significance of the site from a range of environmental perspectives, and identifies potential native remnants for fencing.
3. Development of draft management plans— farmer identifies the remnants and type of fencing and the officer prepares a management plan as the basis for a bid.
4. Submission of bids— farmer submits a sealed bid that declares the amount of payment being sought by her to undertake the agreed management plan.

5. Bid assessment—all bids are assessed objectively on the basis of expected environmental benefits from the fencing activity and the ask price.
6. Winner determination — projects are selected on the basis of 'best-value for money'.
7. Agreements—successful bidders sign agreements based on previously agreed draft management plan.
8. Reporting and payments—periodic payments and reporting occur as specified in the agreement.

The Conservation Reserve Program (CRP) of the USDA is one of the early conservation auction programs (Shoemaker, 1989). Other notable auction programs include BushTender, EcoTender, Auction for Landscape Recovery (ALR) in Australia; and Wetlands Reserve Program (WRP) in USA. These auctions differ in terms of target outcomes and coverage. The key features of some major conservation auction schemes are presented in Table 1. Some conservation auctions (such as the Conservation Reserve Program in the US) have a broad scope and coverage. They may encompass multiple agricultural sectors, different geographic areas and many environmental targets (Reichelderfer and Boggess, 1988). Many farmers (often around 3,000) could participate in such auctions (Vukina et al., 2008). The justifications for having such a wide coverage are to broaden the scope for participation and to improve administrative efficiencies. However, a broadly scoped tender warrants complex auction design and substantial administrative experience. So, conservation auctions are often narrow in conservation goals (e.g. pollution reduction) and confined to specific geographic regions (Rolfe et al., 2008). Key features of the conservation auction designs are elaborated below.

Table 1. Key features of some major conservation auctions

Auction [country]	Objectives & key features	References
BushTender Trial [Australia]	• Objectives: Enhance biodiversity value of remnant or bush • Sealed bid discriminatory price auction • Bids were ranked according to Biodiversity Benefit Index (BBI)	Stoneham et al., 2003
Conservation Reserve Program [USA]	• Objectives: Wildlife enhancement, water quality, erosion control, air quality, and state or national significance. • Farmers submit bids to USDA for individual parcels of land. They receive funds for retiring their lands from farm production for a period of 10 to 15 years. • Bid selection based on Environmental Benefit Index (EBI) • Run several times per year	Latacz-Lohmann and van der Hamsvoort, 1997 Latacz-Lohmann and Schilizzi, 2005 IDNR, 2007
Wetlands Reserve Program (WRP) [USA]	• Objectives: Protect, restore, and enhance wetlands • Applicants submit bids with a self-assessment score • The scores are used to develop an EBI. This EBI was used to rank the applicants for funding. • The applicants are notified of their ranking status and provided an opportunity to submit a lower bid. The lower of the two bids is used to develop a final EBI for funding selection.	USDA, 2008

Table 1. (Continued)

Auction [country]	Objectives & key features	References
EcoTender Multiple-outcome auction of land use change (DPI Victoria) Goulburn-Broken Vic. ID20 [Australia]	• Objectives: Terrestrial biodiversity, water quality, water quantity, salinity reduction and carbon sequestration • First price Sealed Bid, Single round. • Used a Catchment Management Framework Model to develop the Environmental Benefit Index (EBI). Information about Metric Revealed • Bids could be lumped or separate; that is, a farmer could submit a bid for a number of areas or separate bids for each. • Pooled bids across several farmers were also allowed. Payments were annual, subject to satisfactory completion of agreed actions. • Payments were not only input-based (management actions), but also included an output-based element	Grafton, 2005 Latacz-Lohmann and Schilizzi, 2005 Eigenraam et al., 2006
Auction for landscape recovery (WWF Australia) Avon, WA ID21 [Australia]	• Objectives: Diffuse source salinity and biodiversity outcomes • Sealed bid discriminatory price auction • Farmers could put up more than one bid each and were encouraged to put in joint bids. • Tenders were evaluated using a regional metric of 'biodiversity complementarity'. This metric accounts for synergistic aspects due to number, size and distance of several areas.	Grafton, 2005
Establishing East-West Landscape Corridors in the Southern Desert Uplands (Desert Uplands Build-up & Devt Comm.) Burdekin-Fitzroy Qld [Australia]	• Objectives: Establishing landscape corridors • Experimental workshops (with farmers) with multiple rounds of bids. A two-stage process where farmers first bid in terms of individual properties, then this information was subsequently revealed to all participants. Successful at generating landscape scale corridors • In a second round, bidders were allowed to reduce their bid price and / or relocate the location of their own property's conservation area.	Grafton, 2005
Burdekin Water Quality tender [Australia]	• Objective: Water quality improvements across the two industries and three catchment areas • Single bidding round, Sealed bids, Discriminatory pricing • An (unspecified) reserve price, • Multiple bids allowed from farmers, No cap on bids, • One year contracts for successful bidders, Two payment periods for successful bidders: 60% upfront and 40% on completion	Rolfe et al., 2008

Source: Information compiled from the sources.

2.1. Auction Objectives

At present, conservation auctions are used for securing a range of environmental services – from nutrient management to wildlife conservation, often through the same auction. So, generally the bids contain more than one dimension of quality. The inclusion of several targets in a single auctioned project can have benefits (Eigenraam et al., 2006). Firstly, there may be complementarities or jointness in supply or production of environmental goods, e.g., planted trees may simultaneously provide carbon sequestration and wildlife benefits (Baumgartner et al., 2001). Secondly, there could be savings in administrative and monitoring costs. Unlike government procurement tenders for other types of goods, in conservation auctions, farmers submit project proposals where the deliverables are conservation activities (inputs) not the level of environmental services or goods (outcomes). As environmental benefits vary from site to site (non-standard benefits), individual management agreements specify a schedule of management commitments with payments made on the basis of inputs. This allows the farmers to choose from a menu of actions that they prefer (Stoneham et al., 2003). Thus, conservation auctions resemble multi-item procurement auctions, where bids are selected based on established eligibility criteria (Latacz-Lohmann and van der Hamsvoort, 1997).

2.2. Auction Items

Almost all conservation auctions implemented to date have been input-based, where the farmers are paid for undertaking prescribed actions. The widespread practice of input-based vis-à-vis output-based conservation auctions is due to several factors. There is high uncertainty or risk about the production of environmental goods due to the long term and stochastic nature of ecological and environmental processes. There is a lack of understanding of the impact of management actions on the environment, which poses difficulty in linking observed improvement in the environmental conditions with certain actions taken by the participants (Whitten et al., 2004). Because it is hard to link payment with observed environmental changes (Stoneham et al., 2003), governments pay farmers for their efforts instead of results or output.

2.3. Bid Construction

In an auction, a bid is a message that states a participant's willingness to get into a contract or exchange transaction. A bid contains information on an item's specifications and a price or an offer (Wurman et al., 2001). In conservation auctions, initially the conservation agency invites farmers in the target region to submit expression of interests (EoI). Upon receiving the EoI, the agency helps the farmers select a suitable project according to the auction objectives. The bids in conservation auctions usually depend on the forgone benefits, such as the opportunity cost of retaining idle land, which may include foregoing entitled uses, such as firewood collection and grazing rights as well as the cost of active management including labour and material. In the BushTender auction, eligible conservation activities included fencing off remnants, re-vegetation (including planting and direct seeding), feral animal control, weeding and supplementary planting or vegetation (Stoneham et al., 2003). In the ALR, management actions included the fencing off remnants and revegetation, feral animal control, corridor construction, revegetation (including planting and direct seeding), and the institution of nature conservation covenants and voluntary management agreements (Gole et al., 2005). Often farmers are willing to share some costs with the agency and can reduce the ask price in the bid (Rolfe et al., 2008). For example, in the Burdekin Water Quality tender the majority of participants did not include in their bid prices the transaction costs (approximately $220) incurred in developing their bid, and this became part of a de facto cost-share arrangement. Moreover, they were willing to shoulder a portion of the conservation cost, and this ranged from 0% to 95% with a mean of 57% (Greiner et al., 2008).

2.4. Sealed Bid Auction

Conservation auctions follow a sealed bid format, in which bidders submit the bids in sealed envelopes (Krishna, 2002). Sealed-bid auction has some practical advantages of offering flexibility in terms of administration, as it is too costly and time consuming to bring all the bidders (farmers) in the same place at the same time (Wolfstetter, 1996). Also the sealed bid auctions are less susceptible to tacit collusion (Jehiel and Moldovanu, 2003) as bidders are unable to detect and retaliate against bidders who fail to cooperate with them (Abrache et al., 2007). So the government agencies have a long tradition and rich experience in sealed bid procurement tendering.

2.5. Bid Evaluation and Ranking

In conservation auctions, the use of environmental indices (EIs) for bid evaluation and ranking is common (Johansson, and Cattaneo, 2006). Indices are used to generate a common yardstick for project benefits. The EI aggregates various benefit variables into one figure representing an estimate of the overall conservation benefit of each project. The aggregate measure of conservation value draws on environmental assessments of expected benefits (Latacz-Lohmann and Schilizzi, 2005). Winner projects are selected based on the ratio of benefit scores per dollar; this maximizes overall benefits procured by the program (Ferraro, 2004).

Depending on auction objectives, different sets of indices are used for bid selection. In the BushTender, a Biodiversity Benefits Index (BBI) has been used. The BBI depends on Biological Services Score (BSS) and Habitat Services Score (HSS). The BSS draws on information about the scarcity of vegetation types and its Ecological Vegetation Classification, while the HSS represents the change in quality of habitat from management actions. Similar indices have been used in the Victorian Volcanic Plains Tender (Stoneham et al., 2005).

The USDA pioneered the development of Environmental Benefit Indices (EBI) through its Conservation Reserve Program (Lehmann, 2005). Under this system, an EBI is used to indicate the value of different environmental management practices that farmers may implement. The EBI formula considers several environmental factors, such as wildlife, water quality, enduring benefits, air quality, cost, and state or national significance (IDNR, 2007).

Similarly, in the EcoTender, an EBI was used. Considering pre-1750 as the 'natural benchmark' a set of indicators for the following targets are used: terrestrial biodiversity (change in habitat maintained or improved per ha); aquatic function which incorporates changes in water "quality" (tonnes of soil / ha to stream) and water quantity (mm of water / ha to stream); saline land area change (ha with groundwater < 2m); and carbon sequestration (tonnes / ha) (Eigenraam et al., 2006).

In the Auctions for Landscape Recovery (ALR), the EBI was comprised of a Native Biodiversity Benefits Index (NBBI) and an Other Environmental Benefits Index (OEBI). The NBBI utilizes four surrogate measures of biodiversity: vegetation or habitat condition, vegetation or habitat complexity, landscape context, and conservation significance. The OEBI was grouped into two categories – salt, water and soil management benefits, and other environmental benefits or management activities (grazing, fire, weeds and feral animals). The scores from the component attributes were simply summed within each group and then added together to create the OEBI. The final EBI was calculated as the sum of the NBBI and weighted (0.5) OEBI. In the ALR, in addition to EBI, a Strategic Conservation Planning (SCP) tool, which focuses on the complementarities of the sites, was tested for bid evaluation (Gole et al., 2005).

2.6. Discriminatory Pricing

The payments or prices paid to winning bidders do not have to be the same as what the bidders asked. In auction theory (and practice), different pricing rules are identified. The most common are uniform pricing and discriminatory pricing. With uniform pricing, all successful winners are paid the same amount per unit, or environmental benefit in the case of conservation auction. Conservation auctions generally use discriminatory pricing. Under this format, winning bidders receive a payment that is equal to their individual bids or offers (Krishna, 2002). In such cases, a funding agency can reduce costs and gain greater environmental benefits (Gole et al., 2005). However, under discriminatory pricing bidders do not have the incentive to bid truthfully, and may overbid or overstate the true costs of providing the ecological good (Cason and Gangadharan, 2005). To prevent overbidding, the agency may set a reserve price, which is an upper limit on the amount the agency is willing to pay per unit of the conservation good being traded. The agency may pre-announce the reserve price to enhance competition (Stoneham et al., 2000) but this is rarely used in practice.

Compared to alternative allocation methods that provide payment incentives (such as negotiation, fixed payment method) for biodiversity conservation on private lands, auctions are less subjective because bidders rather than agencies determine valuations; and allocations are more transparent because resource allocation is based on an explicit rule for comparing bids (Chan et al., 2003). Through a survey of the participating and non-participating farmers in the ALR auction, Clayton (2005) observed that some of the main reasons behind farmers participation in the auction were – 1) availability of funding support for on-farm activities; 2) flexibility in project formulation compared to fixed payment programs; and 3) 'ownership' or control over the process of project development. Some participants also liked the competitive dimension of the auction, which would ensure best use of money. Similarly, in the case of a

Water Quality tender conducted in Burdekin (Queensland, Australia), Greiner et al. (2008) observed that the farmers were attracted by: the financial incentives, which would enable farmers to conduct investments in conservation practices; the flexibility of being able to submit actions suitable to individual farm circumstances; and technical support provided by the project in considering participation and preparing a submission.

In terms of efficiency and cost minimization, auctions can offer significant advantages (Cason et al., 2003). Stoneham et al. (2003) contended that the BushTender auction mechanism was more effective in minimizing cost when compared to a fixed-price scheme. In the Wetland Reserve Program of the USDA, easement acquisition cost was reported to be reduced by 14 percent when an auction-based mechanism has been adopted. The efficiency and cost minimization of auctions depend on how successfully they can minimize the information rent. In general, conservation auction (bidding process) can reveal the information held by the farmers (opportunity cost of changing land-use), and by the government (ecological value of habitats) in a cost effective manner to identify the most suitable interventions (Latacz-Lohmann and van der Hamsvoort, 1997). Auctions can also provide incentives to farmers to innovate and search for better ways of achieving outputs and attain cost-effectiveness (Rolfe et al., 2008).

3. LIMITATIONS OF EXISTING CONSERVATION AUCTION DESIGNS

So far, conservation auctions have been used to pay the farmers to undertake land management activities to generate mixed environmental and biodiversity benefits. We argue that such 'input based' or 'effort based' auction design is simplistic and arguably inefficient for conserving target species for the following reasons.

3.1. Mixed Objectives

Conservation auctions generally have multiple and mixed environmental and biodiversity objectives, which may range from salinity reduction to native vegetation restoration. Such mixing of objectives can seriously jeopardise actual achievement. The projects, that promise multiple benefits, have higher chances of being selected. The farmers select the interventions, which can generate multiple benefits as per the agency's criteria. For example, the establishment of corridors for reducing isolation in fragmented landscape is a widely accepted intervention in Australian conservation auctions. Hence, the inclusion of corridor establishment in the project proposal will definitely improve the landscape-related score in a score-based index. However, mere establishment of corridors does not ensure reduction of isolation of the patches (Pascual and Perrings, 2007) unless they are designed for a particular species known (or expected) to utilise the corridors (Gol et al., 2005). Another popular intervention is tree plantation or revegetation, which can help with carbon sequestration and habitation for birds. However, the design of plantations should be tailored to the objectives. Some species may prefer dense cover while others may prefer open sparse vegetation. Unless, the targets (which may be conflicting) are taken into consideration, the mere establishment of vegetation may not properly benefit any group of species (Whittingham, 2007).

Also, the emphasis on multi-output interventions means that the projects or interventions targeting specific outcomes have limited chance of either being proposed or selected. Thus, although selected projects can provide some benefits to the target species, they may not be adequate for conservation of those species if the interventions are not sufficient. In Europe, often agri-environment schemes primarily benefit common species rather than endangered and uncommon species of farmland biodiversity (Kleijn et al., 2006). The more successful schemes were either tailored to the needs of a single species (Evans, 1997) or were located near to nature reserves (Peach et al., 2001). Thus, Reid et al. (2007) have contended that 'targeted and evidence-based agri-environment prescriptions are clearly required in order to ensure the realization of species-specific conservation targets'. This observation is appropriate for conservation auctions.

3.2. Lack of Incentive for Innovation

Compared to a fixed payment approach, the existing design of conservation auctions provides the farmer with better flexibility and control of project specification and selection of interventions. However, this flexibility can be improved. In the existing design, interventions are specified when contracts are offered to the farmer and remain fixed for the duration of the contract. Payment depends on farmer's compliance with the contract. This offers the farmers little flexibility and adaptability in response to the changing condition in the project area. An example is the remnant fencing off, which is a common intervention in Australian conservation auctions. Fencing is usually carried out to protect the vegetation from feral animals. However, it may be possible that fencing will prevent other native animals from accessing the vegetation and have a negative effect on local ecology. Moreover, as the farmer is contracted for fencing off the remnant, she will be more concerned about the condition of the fence rather than the condition of the remnant, as her payment depends on the condition of the former. Thus, the existing design provides little incentive and less scope for innovation by the farmers in searching for alternative and more efficient interventions during execution of the project (Whitten et al., 2004).

3.3. Assumption of Risks

In the conventional conservation auction design, the farmers construct their bids based on interventions and the agency select the most cost effective projects with the likelihood of producing multiple benefits. The contracts are based on interventions not on outputs. So the farmers do not assume the risk of producing environmental goods. In the auctions, payment for each intervention is determined through competition and success will depend on cost-benefit ratio of the project. Though the farmers know the cost of the interventions, the agency has better information on the environmental benefits of the interventions (Cason and Gangadharan, 2004). So, given the information asymmetry, finite resource endowment and profit maximizing condition the farmer will choose the interventions that would minimize cost per unit of input, while ignoring the uncertainty of such interventions that benefit any particular species. The existing design does not provide incentive for the farmers to manage the risks. As a consequence, the agency is accepting the risk of producing environmental goods though it does not have any control over or ways to manage the risk.

3.4. Dependence on Scoring System

Dependence on scoring system to select project could be subjective and controversial. The relative position of different bids depend on the information contained in them and the weighting mechanism. Measurement of stock and flow of intrinsically different kinds of environmental goods are compressed into a single conservation benefit score. The benefit scores are thus highly dependent on the relative weights attached to its components. If the weights are changed, the scores will be changed, and so will be the list of successful projects (Hajkowicz et al., 2007). Interventions, which are not covered under existing matrices, have limited chances of being selected. For example, in the first round of the ALR, it was observed that farmers proposed diverse activities ranging from salinity-mitigation through large-scale surface drainage works to site-specific habitat protection programs through local revegetation and fencing activities. Standard benefit assessment protocol for many of the interventions was not developed. So, scoring depended on available data and expert opinion, which was highly variable and subjective. In the second round the matrix was revised (Gol et al., 2005). An alternative scoring system is Data Envelope Analysis (DEA), in which a best practice or efficiency frontier for the submitted bids is constructed. This frontier identifies the intervention combinations that dominate other intervention packages (or projects). Bids that lie on the frontier are selected, as they offer the best value for money (Latacz-Lohmann and Schilizzi, 2005). Another alternative is to use output based auctions, where contracts are made to deliver outputs or results.

3.5. Lack of Flexibility in Project Formulation

In most of the conservation auctions, the farmers are allowed to submit a single project. The project selection is of an 'all or nothing' nature where the farmer is either contracted to carry out the whole project or nothing. However, such project formulation and selection precludes contracting the farmers to perform a number of interventions that are relatively cost-effective (Chan et al., 2003). Moreover, lumpy bids conceal variations in the marginal cost over different combinations of conservation interventions. For example, baiting foxes at a certain point in time may well be more or less costly and have a different effect on an endangered species than baiting foxes elsewhere and at another time. Hence, various combinations of conservation measures or areas may exist that are all able to achieve the conservation aim, albeit at different costs. So, more flexibility could be offered to the farmers in project formulation by allowing them to submit multiple projects with different combinations of project interventions (Wätzold and Drechsler, 2005).

4. OUTPUT-BASED AUCTION

The adoption of output-based auctions is a way of improving the outcomes of conservation auctions. Contrary to the input-based payments, output based payment specifies contracts with farmers defined by environmental outcomes rather than management actions (Grafton, 2005). The farmers are not rewarded for particular actions but for the results of ecological services (Groth, 2005). The winning projects may receive part of their cost when the contracts are assigned and the remainder at the end of the contract period depending on outcomes (Latacz-Lohmann and van der Hamsvoort, 1997). For example, in Germany under a pilot scheme run by the Georg-August-University Goettingen payments were linked to the presence of targeted density of desired plants (endangered grass) in the field over the contract period (Groth, 2008). In the Netherlands, in one agri-environment scheme payment was made on the basis of number of clutches of wader birds present in the field (Verhulst et al., 2007). Similarly, in the Murray Catchment of New South Wales of Australia a project is underway, that links payment with the presence of three ground nesting bird species: Brolga, Bush stonecurlew and Plains wanderer (Gorddard et al., 2008).

The items (biodiversity outcomes) put under the output-based auctions are clearly defined. As noted earlier, a farmer was not be paid for creating a suitable habitat for an endangered plant but only for the actual presence of the plant on his fields or improvement in status of the plant over the contract period (Groth, 2008). The agency has more control and accountability in delivering environmental benefits to the society. It should be noted that the farmers are not required to disclose the information on the interventions, and hence there is less demand for information disclosure in output based auctions. This can reduce monitoring and transaction costs.

This type of auction can provide an incentive for the farmers to be innovative. As the payment is linked to the 'results', farmers will try to produce the results in most cost-effective way. Conservation of species involves both passive (setting the land aside) and active (plantation, predator control) management actions. The options for active management actions can be many. Their effects depend on land-use history, timing of interventions, interactions with other interventions and the response of the target species. Farmers may have private knowledge about the condition of the species on their land and they can devise innovative measures to improve the condition of target species if proper incentives are provided (Gorddard et al., 2008).

In the output-based auctions, the risk of producing the outputs is assumed by the farmers that will depend on the farmer's risk perception. To a risk-neutral farmer, risk is irrelevant, what matters is the expected value of the contract and as a consequence risk-neutral farmers do not require a risk premium to accept a contract. A risk-neutral farmer will accept the contact if the value exceeds the expected value of the best alternative. Risk-averse farmers will only accept a risky alternative if they are paid a risk premium above the expected value of the proposition. Thus, output-based payment can discourage the participation of the risk-averse farmers.

However, output-based payment can offer more flexibility in risk management. By using a dynamic decision framework, farmer can adopt a sequence of conservation investment decisions through time to manage the risk. The timing of investment decisions can significantly impact the ultimate conservation portfolio and a dynamic approach can address how to prioritise the sequence of conservation investments (Ferraro, 2004). Thus, output-based auctions give more control over the environmental outcomes.

Despite the limited information, the assumption is that output-based auctions will be more focused and target oriented, and hence able to achieve the target outcome in a cost effective manner. However, outcome-based approaches require a detailed understanding of the causes and effects of land use practices, and must also take account for natural variations that may be independent of the actions of the farmers' (Grafton, 2005). Farmers need to clearly understand the link between effort and biodiversity outputs. So, the agency must be more specific with its conservation goal and more cautious in selecting projects. Output-based auctions may be suitable for conservation of endangered species. This issue is elaborated in the following sections.

5. CONSERVATION OF MULTIPLE SPECIES

Often the target of present-day environmental policies and programs is to conserve a single species, or an assemblage of species, such as native species, focal or umbrella species, keystone species, endangered or vulnerable species, rare species and nationally important species. In the BushTender, contracts were made with the farmers to maintain existing native vegetation on their farms, whilst in the EcoTender, the target was to achieve biodiversity distribution that was prevalent before 1750 (Strappazzon et al., 2003). In Europe, under the Common Agricultural Policy (CAP), conservation schemes aim at restoring endangered flora and fauna on private lands. However, due to highly complex ecological inter-dependence among the species, it is often required to maintain a variety of species over a landscape. Several species can depend on a single habitat or a particular host, e.g. Elmes and Thomas (1992) have shown that five species of endangered *Maculinea* butterflies in Europe critically depend on a single host *Myrmica* species. So maintenance of this single host species can largely benefit five species of butterflies. Similarly, the Environmental Stewardship Scheme of UK found that one prescribed management intervention benefited an array of species (Table 2). It was more convenient and less costly for the farmers to maintain populations of aerial insects and ground/foliar invertebrates at the same time through a single type of intervention like hedgerow management (Pekec and Rothkopf, 2003).

Table 2. Multiple sets of species benefited from specified interventions under the Environmental Stewardship Scheme of UK (Donald and Evans, 2006)

Management prescription	Aerial insects	Ground / foliar invertebrates	Aquatic invertebrates	Soil-dwelling invertebrates
Hedgerow planting/restoration	√	√		
Ditch management/restoration	√		√	√
Pond and scrape creation/restoration	√		√	√
Water level management				√
Grass strip/margin creation in arable fields		√		
Reduced pesticide/fertilizer inputs	√	√	√	
Wild bird seed mix		√		
Pollen/nectar mix	√	√		
Summer fallows	√	√	√	

From farmers' perspectives, the protection and conservation of species can be interrelated (complementary or substitutable). Complementarities occur when maintenance cost of a bundle of species is lower than the sum of costs required to maintain an individual species of the bundle (Ausubel and Milgrom, 2002). In other words, if A and B are two species, and c (.) denotes the farmer's maintenance cost, A and B are said to be complementary if c ({A, B}) < c ({A}) + c ({B}), and substitutable if c ({A, B}) > c ({A}) + c ({B}) (Abrache et al., 2007). Highly competitive species involve a trade-off such that more of one cannot be conserved without less of the other (Chambers, 1988). In conservation, two species are competitive (complementary) when an increase in conservation effort (x) for one species (s_1) reduces (increases) the probability for conservation of another species (s_2). The equations below illustrate this as the conservation of species $(s_1$ and $s_2)$ depends on the inputs, as well as on the other species (s) (Romstad et al., 2000):

complement   substitute

$$s_1 = f(s_2, x_1), \frac{ds_1}{ds_2} > 0 \qquad\qquad s_1 = f(s_2, x_1), \frac{ds_1}{ds_2} < 0$$

$$s_2 = f(s_1, x_2), \frac{ds_2}{ds_1} > 0 \qquad\qquad s_2 = f(s_1, x_2), \frac{ds_2}{ds_1} < 0$$

$$x = x_1 + x_2 : input \qquad\qquad x = x_1 + x_2 : input \tag{2}$$

The conservation of multiple species can be viewed as a joint production process (Wossink and Swinton, 2007). Jointness in production comes from either technical interdependencies (e.g., conservation of malleefowl positively affects the condition of phascogale and python) (Shumway et al., 1984) or the presence of "non-allocable" inputs (e.g. tree planting affect both malleefowl and phascogale) that cannot be separately managed between products (Havlík et al., 2005). Jointness can also be caused by an allocable fixed input (Wossink and Swinton, 2007). In the case of competitive species, use of more habitats by one species implies fewer habitats for other species (Peerlings and Polman, 2004). However, that two species have jointness does not necessarily mean that they must be conserved in fixed proportions. Alternative technologies typically permit two species to be conserved in different proportions, depending on both the conservation technique and the interventions chosen. For a profit-maximizing producer, the optimal choice of species mix depends on the direct costs of conservation as well as the opportunity cost of not conserving more of the other species (Havlík et al., 2005). The concept of technical and economic complementarities and technical jointness in conservation of multiple-species can be better explained using a multi-species bio-economic model.

6. A BIO-ECONOMIC MODEL ON CONSERVATION OF MULTIPLE SPECIES

The model focuses on three threatened native animals (red-tailed phascogale (RP), carpet python (CP) and malleefowl (MF)) and three pest species (fox, feral cat and rabbit) found in the wheat-belt of Western Australia (Lindenmayer et al., 2003). The objective of the model is to simulate a cost minimizing farmers' conservation activities and bid formulation for cases where the farmer is participating in an output-based auction. It is a model of the farmer's decision-making process. By using this model, we can simulate the performance of output-based auctions and compare different designs. The model meets two criteria: (1) accounting for the effects of management interventions on species conservation at the farm level; and (2) identification of the pattern of management activities on the farm that minimize the farmers' cost over a specified time horizon.

The carpet python (*Morelia imbricata*) is a large (averaging two meters in length and 5 kg in weight), non-venomous, thickset snake (Cogger, 2000). It inhabits woodland with abundant large hollow-bearing trees and/or isolated rocky outcrops. Its diet includes rats, bandicoots, bats, birds and poultry. It requires access to dense cover where it can hide to prey, to avoid predators, and to brood eggs. In suitable environments its density could reach above six snakes per hectare (Shine and Fitzgerald, 1996). Recent studies have shown that the introduced red fox (*Vulpes vulpes*) can be an important predator of the python. The feral cat (*Felis cattus*), tree goanna (*Varanus varius*), and possibly the feral pig (*Sus scrofus*) are other predators (Lindenmayer et al., 2003).

The red-tailed phascogale (*Phascogale calura*) is a small, nocturnal, arboreal, carnivorous marsupial that was once widespread across Australia. The diet of a red-tailed phascogale consists of nectars, insects, small mammals (notably the house mouse (*Mus musculus*)), and birds. Nests, which are constructed of leaves and twigs, are primarily found in and around trees. Population density can be around 1-2 individual / ha (Rhind, 2003). It has been classified as endangered species by both the IUCN Red List and Australian Environment Protection and Biodiversity Conservation Act (EPBC), 1999.

Malleefowl (*Leipoa ocellata*) is a large (about the size of a domestic chicken) ground dwelling bird. It is found in semi-arid to arid shrublands and low woodlands, especially those dominated by mallee (*Eucalyptus* spp.) and/or acacias. A sandy substrate and abundance of leaf litter are required for breeding. In suitable condition one to two pairs could be found per sq km (Booth, 1987). Malleefowl is listed as vulnerable on the Australian Environment Protection and Biodiversity Conservation Act 1999. Natural predators of malleefowl and phascogale include birds of prey, reptiles (e.g. *Varanus gouldii*) and the carpet python, and introduced predators include fox and feral cats.

In our model, all three target species (malleefowl, red-tailed phascogale and carpet python) and the three pest species (fox, feral cat and rabbit) share similar type of habitat though they have different types of interactions with each other. Malleefowl has a competing relationship with rabbit. Phascogale benefits from the nests of the malleefowl, as they are a source of food (insects) for the former. Predators (such as python, feral cat and fox) prey on rabbit, phascogale and malleefowl. Python benefits from the increase in numbers of phascogale and malleefowl. Fox and feral cat are competitors to as well as predators of the carpet python (Molsher, 1999). So, all three of the target species will benefit from management actions that secure habitat; reduce grazing pressure, and fox abundance. The farmer can decide on the interventions depending on the chosen target. For our model, we have selected following types of management actions, which could be practiced on the remnant (for details see Appendix 1) –

1. Retention of remnants: The farmer has to decide the amount of remnants to be put under the conservation scheme.

2. Revegetation / in-planting with native species: In order to improve the quality of the remnant under the scheme revegetation through seeding or planting or both can be undertaken.

3. Weeding / removal of non-native species: Reducing competition to the existing native vegetation through weeding can occur.

4. Stock exclusion: Removal of stock (sheep) from the remnant would protect the remnant vegetation and enhance natural regeneration.

5. Rabbit control: As the rabbit is competitor to the malleefowl, rabbit control will have a positive effect on the malleefowl, though it may affect carpet python and foxes negatively as it is a food source for them.

6. Fox control: Predation by fox is a major cause of mortality of malleefowl and to a lesser extent of phascogale and python (Priddel and Wheeler, 1997). The farmer can adopt baiting, shooting or trapping to control fox population, though baiting is most commonly practiced (Greentree et al., 2000).

7. Feral cat control: Farmer can reduce mortality of malleefowl and phascogale, and competition for python by controlling feral cat population.

8. Asynchronous cropping: By asynchronous cropping the farmer can provide an additional food source to the malleefowl and improve the carrying capacity of remnant.

9. Install nests for phascogale: If there is not adequate number of hollow-bearing trees in the remnant the farmer can install artificial nest boxes for phascogale.

10. Shelter for python: Through leaving logs, litter or shrub cover on the ground; the farmer can increase the carrying capacity of the remnant for python. The farmer can also create artificial ground-layer habitats for python.

11. Relocation of python: Subject to appropriate government permission the farmer can relocate carpet python from the project area.

12. Introduction of malleefowl: Farmer can relocate malleefowl in the project area to increase the existing population of malleefowl.

13. Introduction of phascogale: Similarly, some phascogale can be relocated in to the project area.

We have assumed the size of square remnant is 50 ha and the farmer can decide on the area to put under conservation and interventions. The details of the model (framework, interactions among the species, cost and effect of interventions) are presented in Appendix 1.

6.1. Structure of the Model

The goal of the model is to provide a mechanism for generating cost minimizing activities for the conservation of individual or combinations of species. Let x be a vector or portfolio of conservation activities, w a vector of conservation activity costs, δ the discount rate, and y a population objective. The model optimizes the present value of the cost of conservation. Formally –

$$Min \sum_{t=1}^{t} \frac{C_t(x_j, w)}{(1+\delta)^t}$$

$$s.t.$$

$$f(x_{j,t}) \geq y$$

$$x_{t,t} \geq 0$$

(3)

The biological response function f (·) maps conservation activities at species population levels. The population target, y, is interpreted here as the population level to be achieved at the end of a specified period. Solving the minimization problem yields the set x^*, the vector of conservation activities that minimize the net present value of the cost of satisfying the species population objective y (Rashford and Adams, 2007).

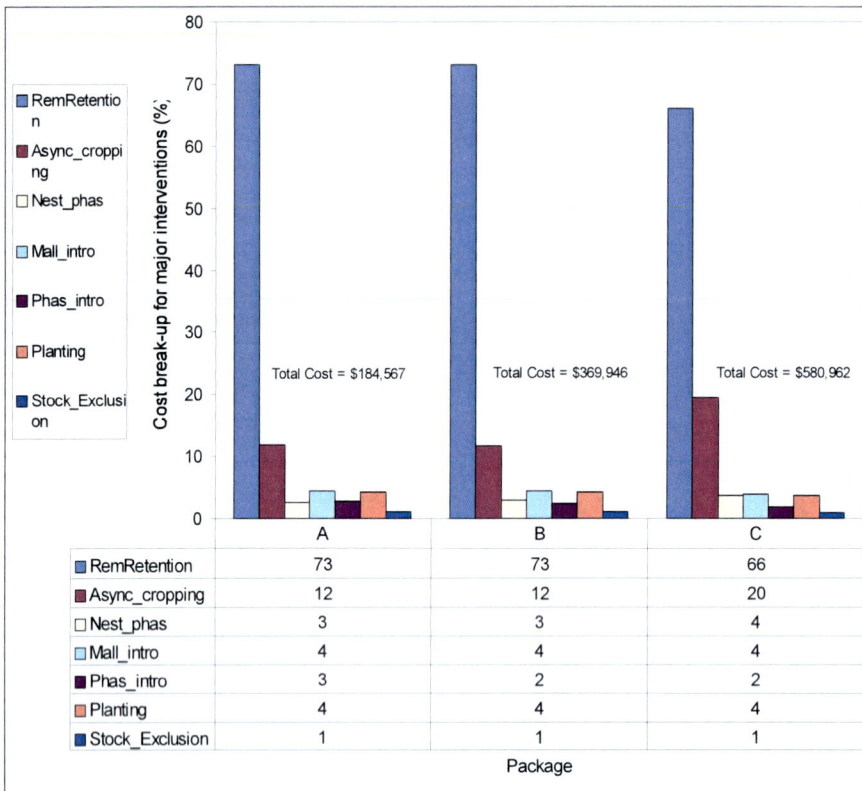

	A	B	C
■ RemRetention	73	73	66
■ Async_cropping	12	12	20
☐ Nest_phas	3	3	4
☐ Mall_intro	4	4	4
■ Phas_intro	3	2	2
■ Planting	4	4	4
■ Stock_Exclusion	1	1	1

Package

Source: The bio-economic model was run to get cost estimates for different population targets (packages).

Figure 1. Costs and major interventions for ten year period obtaining three outcomes packages: A (20 malleefowl, 20 phascogale and 2 pythons), B (40 malleefowl, 40 phascogale and 4 pythons) and C (60 malleefowl, 60 phascogale and 6 pythons) in the final / tenth year

As an example, let's assume that the farmer is interested to know the costs and major interventions for three outcomes packages: A (20 malleefowl, 20 phascogale and 2 pythons), B (40 malleefowl, 40 phascogale and 4 pythons) and C (60 malleefowl, 60 phascogale and 6 pythons) in the tenth year. From Figure 1 we can see that retention of remnants, asynchronous cropping, installation of nests for phascogale, introduction of malleefowl, introduction of phascogale, planting and stock exclusion are the major interventions the farmer will have to carry out to achieve the outcomes.

The model selects the most cost-effective interventions to minimize cost to obtain the packages. So, proportionate cost share of the interventions changes for higher population targets (in packages B and C). The farmer will have to invest proportionately more in costlier interventions like asynchronous cropping and nest boxes for phascogale to conserve higher population sizes. As a result, it becomes costlier to conserve packages of higher population target.

In the following sub-sections we present the cost estimates for achieving different population targets when the target is individual species or a group of species. We have simulated the model on different levels of target (single species and group of species over a range) and traced the changes in population size and conservation cost. Following Schroeder (1992) we investigate two measures of output mix: 1) economies of scope and 2) species-specific and multi-species scale economies to demonstrate the degree to which cost-effectiveness can be improved by allowing the farmers to tender for a combination or a package of conservation outcomes.

6.2. Economies of Scope

The model is flexible enough to allow the farmer to identify interventions and calculate costs for conserving any species individually if she is interested. For example, if the farmer is interested in conserving malleefowl only, the model can estimate conservation costs at different levels of the target. From Figure 2, we can see that the retention of remnants, asynchronous cropping, introduction of malleefowl, relocation of python, planting and stock exclusion are the major interventions the farmer will have to carry out to obtain different sizes of malleefowl population. Average cost for conservation of one malleefowl over ten years period is AUS$ 8,834 (± $601). On the other hand, the farmer will have to carry out interventions like installation of nests for phascogale, weeding, planting, and stock exclusion along with retention of remnants in order to conserve phascogale population of different sizes. Average cost for conservation of one phascogale over ten years period is AUS$ 6,077 (± $519). In case of python conservation, remnant retention, providing shelter for the python, planting and remnant retention are the main interventions. Average cost for conservation of one python over ten years period is AUS$ 41,206 (± $143). There are some common activities such as planting and stock exclusion, which benefit all three species. Fox and feral cat control are common in case of malleefowl and python conservation (Figure 2). This indicates that the farmer can save some cost by conserving multiple species simultaneously (Wossink and Swinton, 2007; Strappazzon et al., 2003).

From the farmer's as well as agency's perspective it is important to see if there is economies of scope (EOS) in conservation of multiple species simultaneously. Whenever the costs of conservation of two or more species are sub-additive (i.e., less than the total costs of providing conserving these species separately), the multi-species cost function exhibits economies of scope (Panzar and Willig, 1981). In other words, in presence of economies of scope the average total cost of conservation decreases as a result of increasing the number of different species conserved. There are economies of scope when EOS is greater than zero. In case of jointness the specialised cost function could be viewed as the cost of producing both goods and then discarding one of them. In our case, following Strappazzon et al. (2003), we can get the specialised cost of conserving one species by simulating the model targeting one species, while ignoring, or 'discarding', the benefits to other two species. Formally,

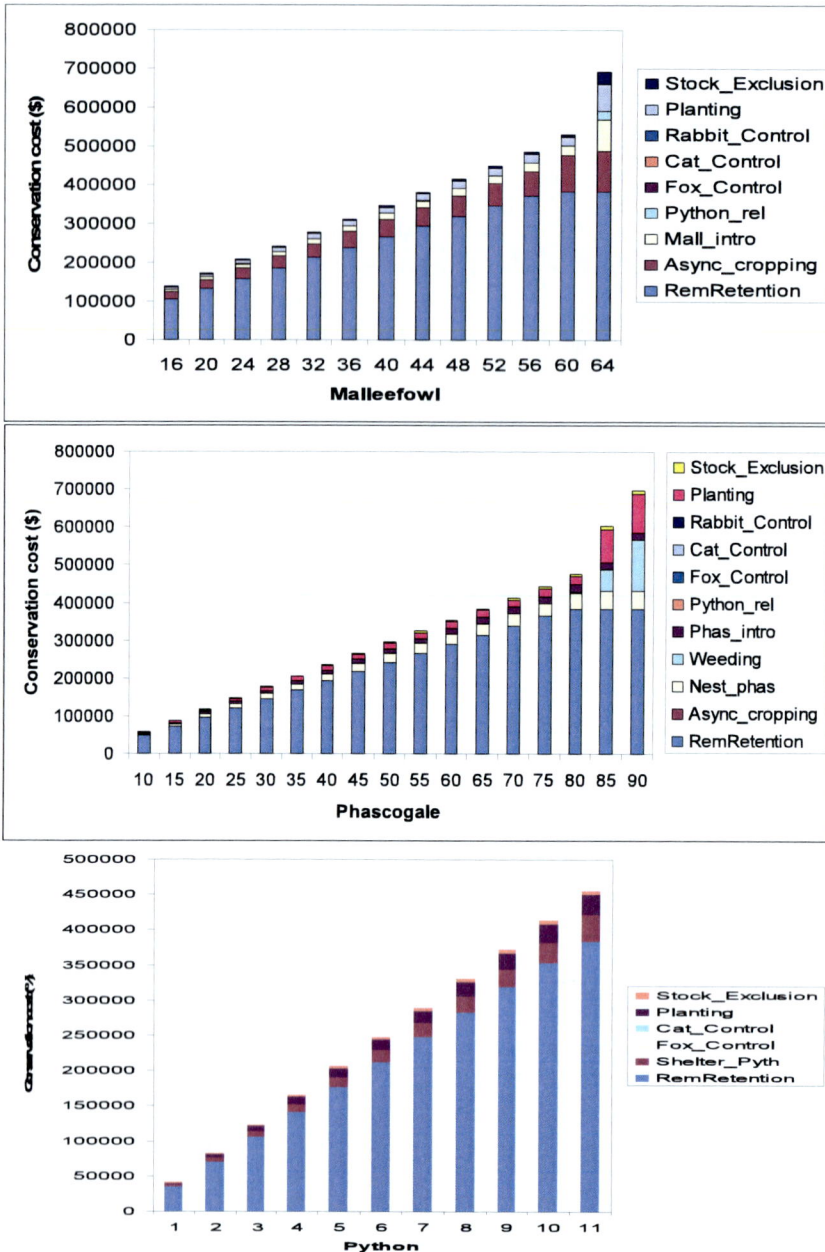

Source: The bio-economic model was run to get cost estimates for different population targets (packages).

Figure 2. Trend in conservation cost and the main interventions when the farmer is targeting different population sizes in the tenth year for individual species

$$EOS_{s^i,s^j} = \frac{c(y_{s^i}) + c(y_{s^j}) - c(y_{s^i}, y_{s^i})}{c(y_{s^i}, y_{s^i})} \times 100$$

where

s^i – *group of species*

s^j – *group of species*

y – *population size*

$i \neq j$

(4)

In order to calculate the EOS, we first estimated conservation costs over a range for the target species individually and as a group. Then we calculated the EOS using equation (4). Figure 3 presents the economies of scope for select packages of species. It can be observed that economies of scope could range from 15% to 140%. In our model, the cost effectiveness comes from ecological interactions and economic complementarities.

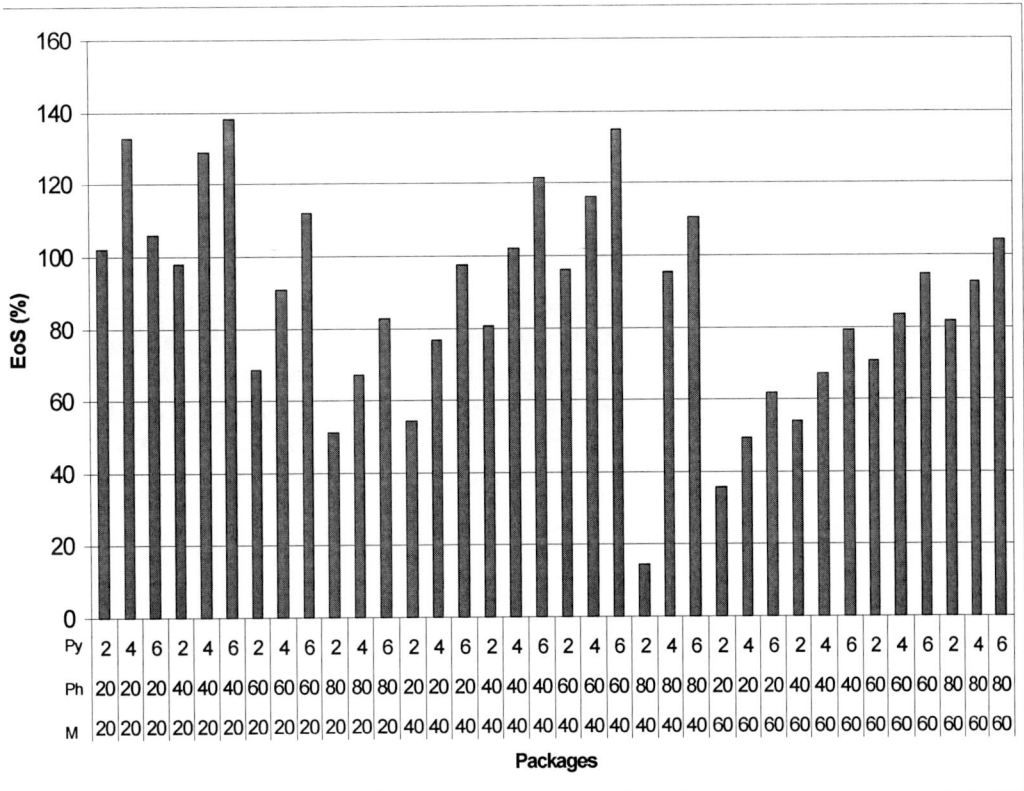

Here, Py = python, Ph = phascogale and M = malleefowl. Packages indicate population sizes at the end of tenth year the farmer is interested to obtain. Source: The bio-economic model was run to get cost estimates for different population targets (packages).

Figure 3. Trend in economies of scope for select outcome packages when the farmer is targeting different population sizes all three species to obtain in the tenth or final year

Phascogale benefits from the presence of malleefowl. Python benefits from presence of both phascogale and malleefowl, though it affects them negatively. Remnant retention is the major source of economic complementarities. Some other interventions such as planting and stock exclusion also affect all target species positively. However, some interventions (such as rabbit control) can benefit one species (malleefowl) and harm another species (python). So, even though the farmer can save a substantial amount of cost by conserving multiple species at the same time care is needed in selecting packages to achieve optimum cost-effectiveness.

6.3. Economies of Scale

Similar to economies of scope, economies of scale refers to the cost advantages that a farm obtains due to expansion. A farm enjoys economies of scale if there is a reduction in long-run average and marginal costs. Economics of scale can be internal to a farm (cost reduction due to technological and management factors) or external (cost reduction due to the effect of technology in an industry). There are two typical ways to achieve economies of scale: 1) High fixed cost and constant marginal cost and 2) Low or no fixed cost and declining marginal cost.

In first case, the initial investment of capital is diffused (spread) over an increasing number of units of output. In second case, the marginal cost of producing a good or service decreases as production increases. Scale economies or returns to scale are usually defined in terms of the relative increase in output resulting from a proportionate increase in all inputs. Brown and Chachere (1980) indicate that it is more appropriate to represent scale economies by the relationship between cost and output along the expansion path where input prices are held fixed and costs are minimized at every level of output (Kim, 1987).

In our case, elasticity of conservation could be defined as the ratio of the proportionate change in cost with respect to proportional change in population size of one species. Given a conservation cost function $C = f(y_s)$ for a single species, where y_s is population size of species s and C respective conservation cost, elasticity is defined as $\varepsilon_{C,y_s} = |\partial \ln C / \partial \ln y_s| = |\partial C / \partial y_s \cdot y_s / C|$. In the case of multi-species packages, where population sizes of all species are changing simultaneously, species specific elasticity could be measured using the following equation –

$$C = a + b_1 M + b_2 Ph + b_3 Py + b_{11} M^2 + b_{22} Ph^2 + b_{33} Py^2 + b_{12} MPh + b_{13} MPy + b_{23} PhPy + b_{123} MPhPy$$

and
$$(5)$$

$$\varepsilon_{C,M\,Ph,Py} = \frac{\partial C}{\partial M} \cdot \frac{M}{C} = (b_1 + 2b_{11}M + b_{12}Ph + b_{13}Py + b_{123}PhPy) \cdot \frac{M}{C}$$

$$\varepsilon_{C,Ph\,M,Py} = \frac{\partial C}{\partial Ph} \cdot \frac{Ph}{C} = (b_2 + 2b_{22}Ph + b_{12}M + b_{23}Py + b_{123}MPy) \cdot \frac{Ph}{C}$$

$$\varepsilon_{C,Py\,M,Ph} = \frac{\partial C}{\partial Py} \cdot \frac{Py}{C} = (b_3 + 2b_{33}Py + b_{13}M + b_{23}Ph + b_{123}MPh) \cdot \frac{Py}{C}$$

Here, *M, Ph* and *Py* are number of malleefowl, phascogale and python respectively. In Figure 4, the elasticity measures for different levels of population sizes when the farmer is targeting individual species are plotted.

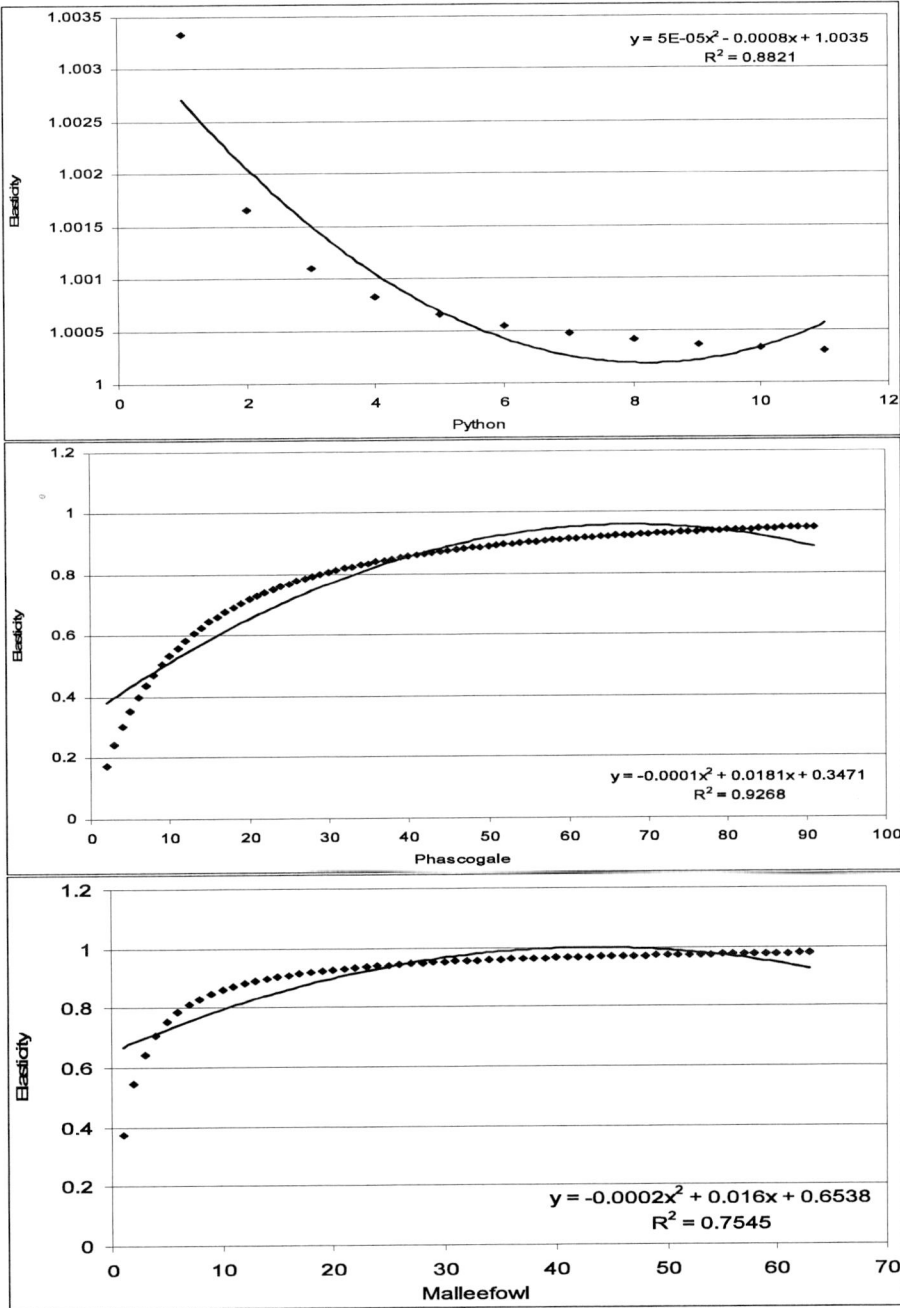

Species-specific elasticities are plotted against targets of different population sizes of individual species the farmer is interested to achieve at the end of tenth year. Source: The bio-economic model was run to get cost estimates for different population targets.

Figure 4. Elasticity measures when the target is an individual species

It becomes more expensive to conserve higher population sizes of python, probably due to the fact that conservation of python depends highly on the amount of land put under conservation and the cost of remnant retention increases with the increase in project area. So, the higher the target, the more land is under conservation and higher the per unit price of land. In cases of phascogale and malleefowl, though species – specific elasticities are less than 1 over the feasible range of population targets, the conservation cost is becoming more sensitive to changes in population sizes at higher targets. It may be more appropriate for the farmer to target moderate population size when individual species are targeted. This becomes clearer if we look into the multi-species scale economies for different packages. As the elasticity measure is ≤ 1 for conservation of packages up to a population size of 20 malleefowl, 60 phacogale and 4 pythons, the conservation cost would be proportionately higher if the farmer targets higher population sizes for any species (Figure 5).

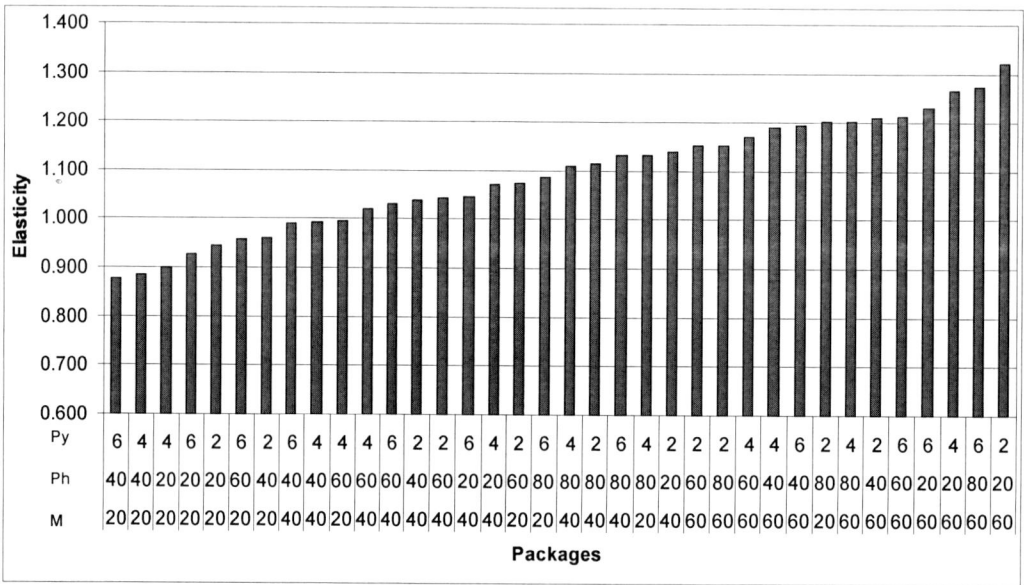

Here, Py = python, Ph = phascogale and M = malleefowl. Packages indicate population sizes at the end of tenth year the farmer is interested to obtain. Source: The bio-economic model was run to get cost estimates for different population targets (packages).

Figure 5. Elasticity measures for select packages when the target is all species

7. COMBINATORIAL BIODIVERSITY AUCTION

In the previous section, we have observed that there is substantial economies of scope (cost saving) when the farmer targets multiple species for conservation. The complementarities arise from the technical jointness in conservation of the species. Combinatorial auctions can handle the cost complementarities better than other designs. It allows simultaneous trading of multiple items. For example, if items *a* and *b* are auctioned, a bid on combination {*a*, *b*} will either win both *a* and *b*, or none (Pekeč and Rothkopf, 2003). The advantage is that the bidder can fully express her preferences as these auctions allow the bidders to place bids on combination or packages of items, rather than on just individual items (Cramton et al., 2006). If the goods are complementary, being able to bid on bundles mitigates the exposure problem; as the bidders can express their precise valuations for any collection of items they desire. So, in addition to the allocation and price discovery, combinatorial auctions offer mechanism for optimal bundling discovery (Pekeč and Rothkopf, 2006).

In a combinatorial biodiversity auction, we propose that the agency can declare willingness to maintain a certain set of target species on the private lands of the target region. Borrowing the terminology from the auction literature, the 'goods' are specified in terms of items (species), units (population of species) and quality (composition of population). Following an invitation for participation in an auction, the farmers place bids showing their willingness to maintain a certain set of species and their respective populations. The ask prices should reflect cost complementarities (if there is any) of maintaining a certain set of species on the part of the bidding farmers. The auctioneer then selects the bids according to her pre-defined criteria, one of which could be minimization of the total cost. The number of individuals of the target species at the final year of the contract period will be put as item of auction. In their bids the participants will state how many individuals of which species they are willing to conserve and at what cost. Examples of some hypothetical bids in a combinatorial auction are presented in Table 3. Here, the bidders have specified the packages / bids in terms of the number of animals for each of the species they wanted to conserve and ask price.

Table 3. Examples of bids in a hypothetical combinatorial biodiversity auction

Bid / Package	Bidder's ID	Item (Species and population size in the tenth year)			Ask price ($)
		Malleefowl	Carpet python	Red-tailed phascogale	
1	01	10	10	2	97,542
2	02	40	40	6	374,538
3	01	60	60	6	580,962
4	02	60	70	4	580,180
5	01	60	80	4	607,742

Source: The bio-economic model was run to get cost estimates for different population targets.

To clarify the concept, let's assume that our agency needs to maintain a set of species, G = $\{1, 2,..., g\}$. It specifies number of animals (or units) of each species it wants, $U = \{u_1, u_2,..., u_g\}$, $u_i \in \Re^+$. There are N sellers $\{1, 2,..., n\}$. Each bidder submits a set of asks, $A_i = \{A_{i1}, A_{i2}, ... , A_{im}\}$. An ask is a tuple $A_{ij} = \left\langle \left(\lambda_{ij}^1, \lambda_{ij}^2,..., \lambda_{ij}^g\right), p_{ij}\right\rangle$, where $\lambda_j^k \geq 0$ is the number of units of species k offered by the bid j submitted by bidder i. The ask price is $p_{ij} \geq 0$. The winner determination problem is to find the least expensive set of bids under the constraint that the agency receives all of the target units of the species (Sandholm et al., 2002):

$$\min \sum_{i=1}^{n} \sum_{j=1}^{2^{k|u_g}-1} p_{ij} x_{ij}$$

$$s.t. \sum_{i,j} \lambda_{ij}^k x_{ij} \geq U \tag{6}$$

$$\sum_i x_{ij} \leq 1$$

$$x_{ij} \in \{0,1\}$$

Here, x_{ij} is a binary variable, indicating whether bundle j is awarded to bidder i. The first constraint is the resource requirement constraint and the second constraint reflects the condition that from each bidder at most one bundle is selected. Depending on the objectives the auctioneer can set additional side-constraints in eq. (6).

While there are potential benefits of combinatorial biodiversity auctions, there are also some challenges in actual implementation of the auctions. Firstly, transfer of risks to the farmer means that participation in these auctions will depend on the risk perceptions of the farmers. If the farmers are highly risk averse, the required risk premium will be high, which may actually favour input-based payment. Secondly, specific targeting of species may provide perverse incentives to the farmer to remove or harm other species from the project area, which are not under contract. Thirdly, due to inadequate ecological knowledge, generally, it is not possible to link individual farmer's action with output for all endangered species. So, the agency can use combinatorial biodiversity auctions only for those species for which it is possible to verify the changes in condition in terms of the individual farmer's actions. However, given the relative ease of monitoring biodiversity and linking them with private property, which is especially true for flora and less mobile fauna like insects and amphibians, combinatorial biodiversity auctions can be economically efficient and ecologically more target-oriented. At least, if carefully designed they have the potential to complement the existing broad-scale conservation auction schemes.

8. CONCLUDING REMARKS

Recently auctions have been used to procure agri-environmental goods in the US, the UK and Australia. The structures of these auctions are generally simple. They are all input-based, where farmers are paid differentially for efforts rather than for outcomes (Wätzold and Schwerdtner, 2005). The farmer defines the project over a multivariate set of attributes/ targets. The agency uses indices to assign weights and individual value functions to the relevant attributes / targets, to calculate a value score. In most of the cases the projects are treated as independent, as the inclusion of any one project does not explicitly influence the benefits score of another. Generally there is a budget constraint for the program (Hajkowicz et al., 2007), and the agency tries to minimise administrative and transaction costs, and to ensure that program objectives are delivered without creating (or enhancing) perverse incentives to the farmers (Rolfe et al., 2008).

However, the existing auction design may not be suitable for conservation of endangered species due to limitations, such as: 1) mixed (often conflicting) environmental objectives; 2) lack of incentive for the farmers to be innovative, as the farmers are paid for interventions rather than for results; 3) acceptance of risk of production of environmental goods by the agency even though it does not have any control or ways to manage the risk during execution of the project; 4) dependence on scoring system, which may be controversial; 5) linking payment with inputs or interventions rather than with changes in environmental condition or outputs; and 6) lack of flexibility for the farmers in project formulation and execution. To make the auctions more result oriented, output-based auctions where payments are linked with biodiversity outcomes, have been trialled. However, these auctions are quite recent and there is much scope for improvement in their designs. There is a need to find auction designs that are more suitable for conservation of a group of endangered species, when the costs for conserving them are complementary or sub-additive to the farmers.

In this paper the concept of combinatorial biodiversity auction is introduced. In such auctions, farmers would be able to submit project proposals to conserve multiple species, and the payment will depend on condition of the species. Through a bio-economic model of three endangered species, the substantial amount of cost complementarities in the conservation of multiple species is demonstrated. Cost complementarity arises from ecological interactions or technical (production based) jointness or both. Combinatorial auctions allow the farmers to utilize the cost complementarity advantages during project formulation (bid submission) and submit most suitable projects, which in turn helps the agency to save spending. Additional advantages of combinatorial biodiversity auction include provision of: 1) more control of the agency in delivering conservation benefits to the society; 2) more incentive to the farmer in designing innovative projects; 3) more control in management of risks of conservation; 4) more target oriented, tailor-made and specific programs for endangered species; 5) more flexibility and control of the farmers in executing their projects. Further study on the designs and field-testing is required before actual implementation of combinatorial biodiversity auctions.

ACKNOWLEDGMENT

The study of M. S. Iftekhar is supported by the International Post-graduate Student Research Scholarship and UWA Post-graduate Award Schemes. A. Hailu and R. Lindner are the supervisors of M. S. Iftekhar's research. Information has been collected from Dr. Patrick Smith, Dr. Suzanne Prober, Mr. Blair Parsons of the CSIRO; Dr. Peter Thomson, Dr. Laurie Twigg , Dr. Garry Gray, Mr. Frank D'Emden of the Department of Agriculture & Food, Western Australia; and Ms. Jessica van der Waag and Ms. Sallyann Harvey of the University of Western Australia. At different stages the bio-economic model has been reviewed by Dr. Patrick Smith and Mr. Blair Parsons of the CSIRO. Mr. Jeff Durkin of the School of Agricultural and Resource Economics, UWA has made grammatical and linguistic corrections. The authors wish to acknowledge their contributions. However, the usual disclaimer applies.

REFERENCES

Abrache, J., Crainic, T. G., Gendreau, M., & Rekik, M. (2007). Combinatorial auctions. *Annals of Operations Research*, 153(1), 131-164.

Armstrong, C. W. (2007). A note on the ecological–economic modelling of marine reserves in fisheries. *Ecological Economics*, 62 (2), 242-250.

Arnold, S. M. (2007). Reintroductions of malleefowl on private land in Westonia are an ecologically and economically viable means of conservation. Honours thesis, *School of Agricultural and Resource Economics,* University of Western Australia, Australia. 71 p.

Ausubel, L. M., & Milgrom, P. R. (2002). Ascending auctions with package bidding. *Frontiers of Theoretical Economics*, 1(1), 1-43.

Baumgartner, S., Dyckhoff, H., Faber, M., Proops, J., & Schiller, J. (2001). The concept of joint production and ecological economics. *Ecological Economics*, 36, 365–372.

Boman, M., Bostedt, G., & Persson, J. (2003). The bioeconomics of the spatial distribution of an endangered species: the case of the Swedish wolf population. *Journal of Bioeconomics*, 5(1), 55-74.

Brown, L., & Chachere, G. (1980). A note on the cost elasticity-scale elasticity relation. *Econometrica*, 48(2), 537-538.

Cason, T. N., Gangadharan, L., & Duke, C. (2003). A laboratory study of auctions for reducing non-point source pollution. *Journal of Environmental Economics and Management*, 46(3), 446-471.

Cason, T. N., & Gangadharan, L. (2005). A laboratory comparison of uniform and discriminative price auctions for reducing non-point source pollution. *Land Economics*, 81(1), 51-70.

Chambers, R. G. (1988). *Applied production analysis: a dual approach.* Cambridge University Press. 331 p.

Chan, C., Laplagne, P., & Appels, D. (2003). The role of auctions in allocating public resources. *Productivity Commission Staff Research Paper,* Productivity Commission, Melbourne. 144 p.

Clarke, P. J. (2002). Experiments on tree and shrub establishment in temperate grassy woodlands: Seedling survival. *Austral Ecology*, 27(6), 606-615.

Clayton, H. (2005). Market incentives for biodiversity conservation in a saline-affected landscape: farmer response and feedback. *Paper presented in the Salinity pre-conference workshop 8 February 2005 of the 49th Annual Conference of the Australian Agricultural and Resource Economics Society, Coffs Harbour 9-11 February 2005.* Source: *http://www.aares.info/files/AARES/Conf2005/AARES2005Papers.htm.* Last accessed: 6 August, 2007.

Cogger, H. G. (2000). *Reptiles and Amphibians of Australia.* Sydney: Reed New Holland. 808 p.

Cramton, P., Shoham, Y., & Steinberg, R. (2006). Introduction to combinatorial auctions. In P. Cramton, Y. Shoham, & R. Steinberg (Eds.), *Combinatorial auctions* (pp. 1-13). Cambridge MA: The MIT Press.

Donald, P. F., & Evans, A. D. (2006). Habitat connectivity and matrix restoration: the wider implications of agri-environment schemes. *Journal of Applied Ecology*, 43(2), 209-218.

Dorrough, J., Moll, J., & Crosthwaite, J. (2007). Can intensification of temperate Australian livestock production systems save land for native biodiversity? *Agriculture, Ecosystems and Environment*, 121, 222–232.

Eigenraam, M., Strappazzon, L., Lansdell, N., Ha, A., Beverly, C., & Todd, J. (2006). EcoTender: auction for multiple environmental outcomes. *Project final report, National Action Plan for Salinity and Water Quality, National Market Based Instruments Pilot Program*, Department of Primary Industries, Victoria, Melbourne, 152 p.

Elmes, G. W., & Thomas, J. A. (1992). Complexity of species conservation in managed habitats: interaction between Maculinea butterflies and their ant hosts. *Biodiversity and Conservation*, 1(3), 155-169.

Engeman, R.M., Shwiff, S. A., Constantin, B., Stahl, M., & Smith, H. T. (2002). An economic analysis of predator removal approaches for protecting marine turtle nests at Hobe sound national wildlife refuge. *Ecological Economics*, 42, 469–478.

Evans, A. (1997). The importance of mixed farming for seed-eating birds in the UK. In D. J. Pain, & M. W. Pienkowski (Eds.), *Farming and Birds in Europe* (pp. 331–357). London, UK: Academic Press.

Ferraro, P. J. (2004). Targeting conservation investments in heterogeneous landscapes : a distance function approach and application to watershed management. *American Journal of Agricultural Economics*, 86(4), 905-918.

Flemming, C. M., & Alexander, R. R. (2003). Single-species versus multiple-species models: the economic implications. *Ecological Modelling*, 170(2-3), 203-211.

Gole, C., Burton, M., Williams, K. J., Clayton, H., Faith, D. P., White, B., Huggett, A., & Margules, C. (2005). *Auction for Landscape Recovery: Final report. WWF-Australia*, Sydney, 191 p.

Gorddard, R., Whitten, S., & Reeson, A. (2008). When should biodiversity tenders contract on outcomes? *Paper presented at the 52nd Annual conference of the Australian Agricultural and Resource Economics Society. Canberra 6-8 February 2008.* Source: *http://ageconsearch.umn.edu/bitstream/5979/2/cp08go02.pdf.* last accessed: 30th June 2008.

Gordon, H. (1954). The economic theory of a common-property resource: the fishery. *Journal of Political Economy*, 62, 124–142.

Grafton, R. Q. (2005). Evaluation of round one of the market based instrument pilot program. *National Market Based Instrument Pilot Program, Action Salinity & Water Australia*, 36 p.

Greentree, C., Saunders, G., McLeod, L., & Hone, J. (2000). Lamb predation and fox control in south-eastern Australia. *The Journal of Applied Ecology*, 37(6), 935-943.

Greiner, R., Rolfe, J., Windle, J., & Gregg, D. (2008). Tender results and feedback from ex-poast participant survey, *Using Conservation Tenders for Water Quality Improvements in the Burdekin Research Report 5,* Central Queensland University, Rockhampton.

Grimm, V. (2007). Sequential versus bundle auctions for recurring procurement. *Journal of Economics*, 90(1), 1-27.

Groth, M. (2005). Auctions in an outcome-based payment scheme to reward ecological services in agriculture – An analysis from a transaction cost economics point of view. *Paper presented in 45th Congress of the Regional Science Association in Amsterdam, 23-27th August 2005.* Source: *http://www.feweb.vu.nl/ersa2005/final_papers/180.pdf.* Last accessed: 6 August, 2007.

Groth, M. (2008). An empirical examination of repeated auctions for biodiversity conservation contracts. *University of Lüneburg Working Paper Series in Economics, 78.* Source: *www.leuphana.de/vwl/papers.*

Haight, R. G., Cypher, B., Kelly, P. A., Phillips, S., Possingham, H. P., Ralls, K., Starfield, A. M., White, P. J. & Williams, D. (2002). Optimizing habitat protection using demographic models of population viability. *Conservation Biology,* 16(5), 1386-1397.

Hailu, A., & Schilizzi, S. (2004). Are auctions more efficient than fixed price schemes when bidders learn? *Australian Journal of Management,* 29(2), 147-168.

Hajkowicz, S., Higgins, A., Williams, K., Faith, D. P., & Burton, M. (2007). Optimisation and the selection of conservation contracts. *The Australian Journal of Agricultural and Resource Economics,* 51, 39–56.

Harlen, R., & Priddel, D. (1996). Potential food resources available to Malleefowl *Leipoa ocellata* in marginal mallee lands during drought. *Australian Journal of Ecology,* 21, 418-428.

Havlík, P., Veysset, P., Boisson, J. M., Lherm, M., & Jacquet, F. (2005). Joint production under uncertainty and multifunctionality of agriculture: policy considerations and applied analysis. *European Review of Agricultural Economics,* 32, 489–515.

IDNR. (2007). *The Environmental Benefits Index Formula (EBI). Iowa Department of Natural Resources.* Source: *http://www.iowadnr.com/forestry/crpebi.html.* Last accessed: 6 August, 2007.

Isaac, R. M., & James, D. (2000). Robustness of the incentive compatible combinatorial auction. *Experimental Economics,* 3(1), 31-53.

Jehiel, P., & Moldovanu, B. (2003). An economic perspective on auctions. *Economic Policy,* 18 (36), 269–308.

Johansson, R. C., & Cattaneo, A. (2006). Indices for working land conservation: form affects function. *Review of Agricultural Economics,* 28 (4), 567 – 584.

Kim, H. Y. (1987). Economies of scale in multi-product firms: an empirical analysis. *Economica,* 54(214), 185-206.

Kleijn, D., Baquero, R. A., Clough, Y., Diaz, M., De Esteban, J., Fernández, F., Gabriel, D., Herzog, F., Holzschuch, A., Jöhl, R., Knop, E., Kruess, A., Marshall, E. J. P., Steffan-Dewenter, I., Tscharntke, T., Verhulst, J., West, T. M., & Yela, J. L. (2006). Mixed biodiversity benefits of agri-environment schemes in five European countries. *Ecology Letters,* 9, 243–254.

Krishna, V. (2002). *Auction theory.* San Diego: Academic Press. 303+xi p.

Latacz-Lohmann, U., & Schilizzi, S. (2005). Auctions for conservation contracts: a review of the theoretical & empirical literature. *Report to the Scottish Executive Environment and Rural Affairs Department.* 101 p.

Latacz-Lohmann, U., & Van der Hamsvoort, C. (1997). Auctioning conservation contracts: a theoretical analysis and application. *American Journal of Agricultural Economics,* 79, 407–418.

Lehmann, P. (2005). *An economic evaluation of the U.S. Conservation Reserve Program. UFZ-Discussion Papers, Department of Economics, UFZ Centre for Environmental Research, Leipzig, Germany.* Source: *http://www.ufz.de/data/Disk_Papiere_2005-012519.pdf.* 59 p.

Lindenmayer, D. B., Claridge, A., Hazell, D., Michael, D., Crane, M., MacGregor, C., & Cunningham, R. (2003). Wildlife on farms : how to conserve native animals. *Commonwealth Scientific and Industrial Research Organisation (CSIRO)*. Victoria.

Lindenmayer, D. B., Cunningham, R., Crane, M., Michael, D., & Montague-Drake, R. (2007). Farmland bird responses to intersecting replanted areas. *Landscape Ecology*, 22(10), 1555-1562.

Mack, R. N., Simberloff, D., Lonsdale, W. M., Evans, H., Clout, M., & Bazzaz, F. A. (2000). Biotic invasions: causes, epidemiology, global consequences, and control. *Ecological Applications*, 10(3), 689-710.

Matulich, S. C., & Hanson, J. E. (1986). Modeling supply response in bioeconomic research: an example from wildlife enhancement. *Land Economics*, 62(3), 292-305.

McAfee, R. P., & McMillan, J. (1987). Auctions and bidding. *Journal of Economic Literature*, XXV, 699-738.

Molsher, R. L. (1999). Trapping and demographics of feral cats (*Felis catus*) in central New South Wales. *Wildlife Research*, 28, 631–636.

Müller, K., & Weikard, H. P. (2002). Auction mechanisms for soil and habitat protection programmes. In K. Hagedorn (Ed.), *Environmental Co-operation and Institutional Change* (pp. 200-211). Cheltenham: Edward Elgar.

Musters, C. J. M., Kruk, M., de Graaf, H. J., & Ter Keurs, W. J. (2001). Breeding birds as a farm product. *Conservation Biology*, 15(2), 363-369.

Nalle, D. J., Montgomery, C. A., Arthur, J. L., Polasky, S., & Schumaker, N. H. (2004). Modeling joint production of wildlife and timber in forests. *Journal of Environmental Economics and Management*, 48, 997–1017.

Panzar, J. C., & Willig, R. D. (1981). Economies of scope. *The American Economic Review*, 71(2), 268-272.

Parkes, D. C. (2006). Iterative combinatorial auctions. In P. Cramton, Y. Shoham, & R. Steinberg (Eds.), *Combinatorial auctions* (pp. 41-77). Cambridge MA: The MIT Press.

Pascual, U., & Perrings, C. (2007). Developing incentives and economic mechanisms for in situ biodiversity conservation in agricultural landscapes. *Agriculture, Ecosystems & Environment*, 121(3), 256-268.

Peach, W. J., Lovett, L. J., Wotton, S. R., & Jeffs, C. (2001). Countryside stewardship delivers cirl buntings (*Emberiza cirlus*) in Devon, UK. *Biological Conservation*, 101(3), 361-373.

Peerlings, J., & Polman, N. (2004). Wildlife and landscape services production in Dutch dairy farming; jointness and transaction costs. *European Review of Agricultural Economics*, 31(4), 427-449.

Pekec, A., & Rothkopf, M. H. (2003). Combinatorial auction design. *Management Science*, 49(11), 1485-1503.

Polasky, S., Camm, J. D., & Garber-Yonts, B. (2001). Selecting biological reserves cost-effectively: an application to terrestrial vertebrate conservation in Oregon. *Land Economics*, 77(1), 68-78.

Polasky, S., Camm, J. D., Solow, A. R., Csuti, B., Whitee, D., & Ding, R. (2000). Choosing reserve networks with incomplete species information. *Biological Conservation*, 94(1), 1-10.

Pressey, R. L., Possingham, H. P., & Day, J. R. (1997). Effectiveness of alternative heuristic algorithms for identifying indicative minimum requirements for conservation reserves. *Biological Conservation*, 80(2), 207-219.

Priddel, D., & Wheeler, R. (1997). Efficacy of fox control in reducing the mortality of released captive-reared malleefowl, *Leipoa ocellata. Wildlife Research*, 24, 469–482.

Rashford, B. S., & Adams, R. M. (2007). Improving the cost-effectiveness of ecosystem management: an application to waterfowl production. *American Journal of Agricultural Economics*, 89(3), 755-768.

Reichelderfer, K. R., & Boggess, W. G. (1988). Government decision making and program performance: the case of the conservation reserve program. *American Journal of Agricultural Economics*, 70(1), 1-11.

Reid, N., McDonald, R. A., & Montgomery, W. A. (2007). Mammals and agri-environment schemes: hare haven or pest paradise? *The Journal of applied ecology*, 44(6), 1200-1208.

Rhind, S. G. (2003). Communal nesting in the usually solitary marsupial, *Phascogale tapoatafa. Journal of Zoology*, 261, 345-351.

Risbey, D. A., Calver, M. C., Short, J., Bradley, J. S., & Wright, I. W. (2000). The impact of cat and foxes on small vertebrate fauna of Heirisson Prong, Western Australia. II. A field experiment. *Wildlife Research*, 27, 223–235.

Robley, A., Reddiex, B., Arthur T., Pech R., & Forsyth, D. (2004). Interactions between feral cats, foxes, native carnivores, and rabbits in Australia. *Arthur Rylah Institute for Environmental Research, Department of Sustainability and Environment*, Melbourne. 76 p.

Rohweder, M. R., McKetta, C. W., & Riggs, R. A. (2000). Economic and biological compatibility of timber and wildlife production: an illustrative use of production possibilities frontier. *Wildlife Society Bulletin*, 28(2), 435-447.

Rolfe, J., Griener, R., Windle, J., Hailu, A., & Gregg, D. (2008). *Testing for scope and scale efficiencies in water quality tenders: Final Report. Using Conservation Tenders for Water Quality Improvements in the Burdekin, Research Report 7,* Central Queensland University, Rockhampton.

Romstad, E., Vatn, A., Rørstad, P. K., & Søyland, V. (2000). Multifunctional agriculture: Implications for policy design. *Report No. 21, Agricultural University of Norway, Department of Economics and Social Sciences*. Ås, 158 p.

Rousseau, S., & Moons, E. (2008). The potential of auctioning contracts for conservation policy. *European Journal of Forest Research*, 127(3), 183-194.

Sandholm, T., Suri, S., Gilpin, A., & Levine, D. (2002). Winner determination in combinatorial auction generalizations. In: *International Conference on Autonomous Agents and Multi-Agent Systems (AAMAS)*, Bologna, Italy, July, pp. 69–76.

Scarff, F. R., Rhind, S. G., & Bradley J. S. (1998). Diet and foraging behaviour of brush-tailed phascogales (*Phascogale tapoatafa*) in the jarrah forest of south-western Australia. *Wildlife Research*, 25, 511-526.

Schirmer, J., & Field, J. (2000). *The cost of revegetation*. Final Report. Australian National University. 116 p. Source:
http://www.environment.gov.au/land/publications/costrev/pubs/costrev.pdf.

Schroeder, T. C. (1992). Economies of scale and scope for agricultural supply and marketing cooperatives. *Review of Agricultural Economics*, 14(1), 93-103.

Shine, R., & Fitzgerald, M. (1996). Large snakes in a mosaic rural landscape: The ecology of carpet pythons *Morelia spilota* (serpentes: Pythonidae) in coastal eastern Australia. *Biological Conservation*, 76(2), 113-122.

Shoemaker, R. (1989). Agricultural land values and rents under the conservation reserve program. *Land Economics*, 65(2), 131-137.

Shumway, C. R., Pope, R. D., & Nash, E. K. (1984). Allocatable fixed inputs and jointness in agricultural production: implications for economic modelling. *American Journal of Agricultural Economics*, 66(1), 72-78.

Skonhoft, A. (1999).On the optimal exploitation of terrestrial animal species. *Environmental and Resource Economics*, 13(1), 45–57.

Stoneham, G., Chaudhri, V., Ha, A., & Strappazzon, L. (2003). Auctions for conservation contracts: an empirical examination of Victoria's BushTender trial. *The Australian Journal of Agricultural and Resource Economics*, 47 (4), 477–500.

Stoneham, G., Lansdell, N., Cole, A., & Strappazzon, L. (2005). Reforming resource rent policy: an information economics perspective. *Marine Policy*, 29, 331–338.

Strappazzon, L., Ha, A., Eigenraam, M., Duke, C., & Stoneham, G. (2003). Efficiency of alternative property right allocations when farmers produce multiple environmental goods under the condition of economies of scope. *The Australian Journal of Agricultural and Resource Economics*, 47(1), 1-27.

Swanson, T. (1994). The economics of extinction revisited and revised: a generalised framework for the analysis of the problems of endangered species and biodiversity losses. *Oxford Economics Papers*, 46, 800–821.

Thomson, P. C. (1986). The effectiveness of aerial baiting for the control of dingoes in north-western Australia. *Australian Wildlife Research*, 13, 165–176.

Thomson, P. C., & Algar, D. (2000). The uptake of dried meat baits by foxes and investigations of baiting rates in Western Australia. *Wildlife Research*, 27, 451–456.

Traill, B. J., & Coates, T. D. (1993). Field observations on the brush-tailed phascogale *Phascogale tapoatafa* (Marsupialia : Dasyuridae). *Australian Mammalogy*, 16, 61- 65.

Twigg, L. E., Martin, G. R., & Lowe, T. J. (2002). Evidence of pesticide resistance in medium-sized mammalian pests: a case study with 1080 poison and Australian rabbits. *Journal of Applied Ecology*, 39(4), 549-560.

Twyford, K. L., Humphrey, P. G., Numm, R. P., & Willoughby, L. (2000). Eradication of feral cats (*Felis catus*) from Gabo Island, south-east Victoria. *Ecological Management & Restoration*, 1, 42–49.

USDA. (2008). Wetlands Reserve Program. *Natural Resource Conservation Service, United States Department of Agriculture.* Source: *http://www.nrcs.usda.gov/Programs/WRP/.* Last accessed: 30th June 2008.

Verhulst, J., Kleijn, D., & Berendse, F. (2007). Direct and indirect effects of the most widely implemented Dutch agri-environment schemes on breeding waders. *Journal of Applied Ecology*, 44, 70–80.

Vesk, P. A., & Dorrough, J. W. (2006). Getting trees on farms the easy way? Lessons from a model of eucalypt regeneration on pastures. *Australian Journal of Botany*, 54, 509–519.

Vukina, T., Zheng, X., Marra, M., & Levy, A. (2008). Do farmers value the environment? Evidence from a conservation reserve program auction. *International Journal of Industrial Organization*, 26(6), 1323-1332.

Wätzold, F. & Drechsler, M. (2005). Spatially uniform versus spatially heterogeneous compensation payments for biodiversity-enhancing land-use measures. *Environmental and Resource Economics*, 31(1), 73-93.

Wätzold, F. & Schwerdtner, K. (2005). Why be wasteful when preserving a valuable resource? A review article on the cost-effectiveness of European biodiversity conservation policy. *Biological Conservation*, 123(3), 327-338.

Whitten, S., Carter, M., & Stoneham, G. (2004). Market-based tools for environment management. *Proceedings of the 6th Annual AARES national symposium 2003. RIRDC Publication No 04/142, Rural Industries Research and Development Corporation (RIRDC).* Source: *http://www.rirdc.gov.au.* 216 p.

Whittingham, M. J. (2007). Will agri-environment schemes deliver substantial biodiversity gain, and if not why not? *Journal of Applied Ecology*, 44(1), 1-5.

Wolfstetter, E. (1996). Auctions: an introduction. *Journal of Economic Surveys*, 10(4), 367-420.

Wossink, A., & Swinton, S. M. (2007). Jointness in production and farmers' willingness to supply non-marketed ecosystem services. *Ecological Economics*, 64(2), 297-304.

Wurman, P. R., Wellman, M. P., & Walsh, W. E. (2001). A parametrization of the auction design space. *Games and Economic Behavior*, 35(1-2), 304-338.

Xia, M., Koehler, G. J., & Whinston, A. B. (2004). Pricing combinatorial auctions. *European Journal of Operational Research*, 154(1), 251-270.

APPENDIX -1: A BIO-ECONOMIC MODEL ON ENDANGERED SPECIES CONSERVATION IN THE WHEAT BELTS OF WA

Originating from Gordon's (1954) seminal work, bio-economic models have traditionally focused on fisheries resources. Swanson (1994) applied a model in the context of terrestrial species. Since then, bio-economic models have been used to analyse the impacts of harvesting or extraction of animals, the establishment of protected areas or reserves, the declaration of new legislations, forest management practices and selection of land-use practices (Skonhoft, 1999; Armstrong, 2007). There is a set of bio-economic models that focuses on cost-effective species conservation on farmland. Often the models are used to identify the smallest number or cheapest set of sites to realize targeted wildlife criteria (Polasky et al., 2001). Primarily wildlife criteria have often been the minimization of extinction risk through population viability analysis (Haight et al., 2002) or the coverage of a maximum number of species within a budget constraint based on species distribution (often probabilistic) data (Polasky et al., 2000). Often conservation of farmland biodiversity and agriculture commodity production has been modelled as a joint production process (Wossink and Swinton, 2007), and bio-economic models are used to test compatibility between commodity production and conservation of biodiversity under different land-use practices (Rohweder et al., 2000; Nalle et al., 2004).

Despite some limitations, these and similar models have generated important insights into the behaviour exhibited by the farmers (Flemming and Alexander, 2003). Most of the farm level models have not considered conservation activities that directly manipulate the species through the control of predators and direct habitat manipulation (Rashford and Adams, 2007).

Studies of the cost-effectiveness of direct conservation activities have generally considered single activities, such as harvest (Boman et al., 2003) and predator control (Engeman et al., 2002). Several studies that examine diverse sets of conservation activities have either not included species interactions (Rashford and Adams, 2007) or interaction among the activities (Matulich and Hanson, 1986) in their models. Rarely has the population size of target species been considered as a target in the models.

We are interested in a model that would trace population size of interacting endangered species under different types of conservation interventions in the habitat. We are not aware of any study describing such a model that is applicable in the context of endangered species found in the wheat-belt of Western Australia. Moreover, it is very difficult to adapt any existing model and generalize any population's response to a management action, not only because of the complex biology but also the intricate web of interactions within and outside a habitat with different factors at a multitude of spatial and temporal scales. So, in the absence of an existing model, we have developed a hypothetical bio-economic model. The structure of the model was developed based on existing models (for a review see Robley et al., 2004). Information on cost and baseline condition of the remnant was collected from Commonwealth Scientific and Industrial Research Organisation (CSIRO) and University of Western Australia. Finally, based on expert review, the equations and values of the parameters were refined through iterations. So, results from model should be considered as indicative only.

This bio-economic model traces the dynamics in population size of three endangered target species (malleefowl, red-tailed phascogale and carpet python) and three pest species (fox, feral cat and rabbit) in response to the interventions undertaken by the farmers. The model derives the cost function of the farmer to conserve populations of different sizes of these target species. We have envisioned a remnant, located in the wheat-belt of Western Australia, which is mostly composed of trees suitable for the target species, such as whipstick, bull mallee, *Melaleuca* spp. and *Casuarina* spp., and shrubs like *Maireana* spp. and *Atriplex* spp. (Schirmer and Field, 2000). It has been assumed that due to different declining factors the condition of the target species is very poor. The farmer can take interventions to improve the condition of target species.

The structure of the model is described below.

1.1. Biological Component of the Model

1.1.1. Vegetation Dynamics in the Remnant

In the model, the vegetation has two main components – tree (*tr*) and shrubs (*shr*). Shrubs include grasses and pasture. Trees (*tr*) are divided into functional groups of seed (*tr1*), seedling (*tr2*), sapling (*tr3*) and mature tree (*tr4*). Transition within tree category from one functional group to another group has been modelled as follows –

$$tr1_t = tr4_{t-1} \times 100$$
$$tr2_t = tr1_{t-1} \times 0.2 + tr2_{t-1} \times (0.2 - tr2_{t-1} \times e^{-12.5} - tr3_{t-1} \times e^{-12})$$
$$tr3_t = tr2_{t-1} \times (0.05 - tr2_{t-1} \times e^{-12}) + tr3_{t-1} \times (0.9 - tr3_{t-1} \times e^{-10} - tr4_{t-1} \times e^{-10})$$
$$tr4_t = tr3_{t-1} \times 0.02 + tr4_{t-1} \times 0.95$$

$$(1)$$

Here, $tr1$, $tr2$, $tr3$ and $tr4$ are number of seeds, seedlings, saplings and mature tree per ha. It shows that transition within tree category from one functional group to another group will be gradual and depend on the condition of vegetation on previous year (Clarke, 2002). It has been assumed that the survival rate of seedlings and saplings is density dependent, so that with increasing density the survival rate will be slower. The number of seedlings per ha depends on number of saplings growing into seedling category and number of seedlings survived from last year.

Production of green (forageable) biomass (V) will depend on the number of plants present in the remnant. For trees, we have assumed that forageable green biomass, is produced at a certain proportion depending on the number of trees. For pasture and shrubs, following Vesk and Dorrough (2006), the biomass production depends on previous year's biomass load and winter rainfall. Formally,

$$V(tr1)_t = tr1_t \times 0.002$$
$$V(tr2)_t = tr2_t \times 5$$
$$V(tr3)_t = tr3_t \times 10$$
$$V(tr4)_t = tr4_t \times 30 \qquad (2)$$
$$V(tree)_t = V(tr1)_t + V(tr2)_t + V(tr3)_t + V(tr4)_t$$
$$V(Pas)_t = V(Pas)_{t-1} + V(Pas)_{t-1} \times 0.0002 \times 400 \times \left(1 - \frac{V(Pas)_{t-1}}{6000}\right)$$
$$V_t = V(tree)_t + V(Pas)_t$$

Here, $V(Pas)$ and $V(tree)$ are biomass from pasture and trees respectively in terms of Kg ha^{-1}, V_t is total biomass. $V(tr1)$, $V(tr2)$, $V(tr3)$ and $V(tr4)$ are green biomass produced from seeds, seedlings, saplings and trees respectively. $V(Pas)_{t-1}$ is previous year's pasture load. 6,000 kg ha^{-1} has been considered as the maximum pasture biomass the remnant could support. Average winter rainfall has been assumed as 400 mm. It is assumed from one seed, seedling, sapling and mature tree that respectively 0.002, 5, 10 and 30 kilogram green biomass is available for consumption. Equation (1) and (2) shows the dynamics in vegetation in absence of any animal and farmer's interventions They ignore inter and intra specific competition within the plant community, any changes in pasture composition, and assumes an even spatial distribution of vegetation.

1.1.2. Effects of Rabbits, Sheep and Malleefowl on Vegetation

The consumption of biomass by herbivores has been measured using the equation developed by Short (1987 cited in Robley et al., 2004). The daily per capita consumptions of green biomass by rabbit, malleefowl and sheep adjusted for body weight and expressed as kg animal^{-1}day^{-1} have been estimated as 0.11, 0.13 and 0.49 respectively. It has been assumed that grazing is practiced for four months in the remnant. Consumption of green biomass by the herbivores and influence on the existing vegetation have been modelled using the following equations -

$$cons_t = 0.49 \times sheep_t \times 120 + 0.11 \times rabbit_t \times 365 + 0.13 \times mallee_t \times 365$$

$$V(\text{mod})_t = V_t - cons_t$$

$$cons(Pas)_t = cons_t \times V(Pas)_t / V_t$$

$$cons(tree)_t = cons_t \times V(tree)_t / V_t \qquad\qquad (3)$$

$$V(Pas)_{\text{mod},t} = V(Pas)_t - cons(Pas)_t$$

$$tr1(\text{mod})_t = (cons(tree)_t \times V(tr1)_t / V(tree)_t) / 0.002$$

$$tr2(\text{mod})_t = (cons(tree)_t \times V(tr2)_t / V(tree)_t) / 5$$

$$tr3(\text{mod})_t = (cons(tree)_t \times V(tr3)_t / V(tree)_t) / 10$$

Here, $sheep_t$, $rabbit_t$ and $mallee_t$ are population sizes of respective animals; $cons_t$ is the amount of biomass consumed by the herbivores expressed in kg ha^{-1}. $V(mod)_t$ is the green biomass after consumption. $cons(Pas)$ and $cons(tree)$ are amount of green biomass consumed from pasture and tree category respectively. $V(Pas)_{mod,\,t}$ is the amount of pasture used for calculating next year's pasture load. We have assumed that the animals are indifferent to sources of forageable biomass and consume proportionately to the biomass production from pasture and trees. In order to calculate the effect of grazing on trees we have estimated the rate of consumption for each tree category (seed, seedling and sapling) at a rate proportionate to their contribution in the total green tree biomass. Then we have divided that amount of consumption with the amount of green biomass produced from individuals in each category to get the number of seeds, seedlings and saplings removed from the remnant due to animal consumption. The effect of consumption on mature trees is not very adverse and there is no removal of mature tree due to consumption.

Availability of forageable green biomass (V_t) will determine the carrying capacity of the site for animals. In a study in Victoria, it has been observed that foraging in trees make up 80% - 90% of phascogale activity (Traill and Coates, 1993) and diet consists predominantly of invertebrates and eucalypt nectar when obtainable (Scarff et al., 1998). So, we have assumed that the carrying capacity for phascogale ($K_{phus,\,t}$) depends on the number of trees present in the remnant. On the other hand, malleefowl's diet is primarily seeds, herbs and ground-dwelling invertebrates, which are dependent on the quality of the former two (Harlen and Priddel, 1996). So, we have assumed that the carrying capacity of rabbit and malleefowl ($K_{rabbit,\,t}$, $K_{mallee,\,t}$) depend on forgeable biomass of the remnant. Formally,

$$K_{rabbit,t} = 0.0002 \times V_t$$

$$K_{mallee,t} = 0.00008 \times V_t \qquad\qquad (4)$$

$$K_{phas,t} = tr2_t \times 0.0005 + tr3_t \times 0.0006 + tr4_t \times 0.0008$$

1.1.3. Interactions Among Predators and Prey

To trace interactions among the predator and prey populations we have utilized a variant of Lotka-Volterra equation. Let's assume that prey population sizes are $H = \{h^1, h^2, ..., h^i\}$ and predator population sizes are $P = \{p^1, p^2, ..., p^i\}$. Formally,

$$h_t^j = h_{t-1}^j + r^j h_{t-1}^j \left(1 - \frac{h_{t-1}^j}{K_t^j}\right) - \sum_{j\neq l} \beta_{j,l} h_{t-1}^j h_{t-1}^l - \sum_{p=1}^{i} \beta_{j,i} h_{t-1}^j p_{t-1}^i$$

(5)

$$p_t^i = p_{t-1}^i + r^i p_{t-1}^i \left(1 - \frac{p_{t-1}^i}{\sum_{h=1}^{j} k_j^i h_{t-1}^j}\right) - \sum_{i\neq k} \beta_{i,k} p_{t-1}^i p_{t-1}^k$$

Here, h_t^j is the population size of any prey species in time t, p_t^i is the population size of any predator species in time t, K^j is the carrying capacity of the site for prey j, r^j is growth rate of prey population, r^i is growth rate of the predator population, K^i is the carrying capacity of predator per 1 prey and β is the interaction effect between species. In general, for prey species presence of competing or predator species reduces population size. So without predators, prey population density increases according to a logistic model. Predator dynamics is represented by a logistic model with carrying capacity proportional to the number of prey. For predator species, population size increases in presence of prey species and declines in presence of competing predators. In absence of prey population the predator will become extinct. The biological parameters and their hypothetical values used in the model for simulations are presented in Table 4. Information on baseline condition of the remnant; i.e., initial sheep density and vegetation density; has been collected from CSIRO.

Table 4. Baseline condition of the patch and parameters used in the model

	Malleefowl	Red-tailed phascogale	Carpet python	Rabbit	Fox	Feral cat
Initial population density per ha (h_0 & P_0)	0.0001	0.0004	0.04	4	0.05	0.06
Initial intrinsic growth rate (r)	0.3	0.4	0.2	1.2	0.5	0.6
Interaction co-efficient (β)						
Malleefowl	0	0	(-) 0.04	(-) 0.001	(-) 0.05	(-) 0.02
Phascogale	(+) 0.000005	0	(-) 0.03	0	(-) 0.05	(-) 0.05
Python	0	0	0	0	(-) 0.005	(-) 0.002
Rabbit	(-) 0.001	0	(-) 0.04	0	(-) 0.06	(-) 0.03
Fox	0	0	(-) 0.005	0	0	(-) 0.03
Cat	0	0	(-) 0.002	0	(-) 0.03	0
Carrying capacity of predator per 1 prey (k)						
Carpet python	0.4	0.4		0.4	0.05	0.1
Fox	0.4	0.4	0.2	0.3		0.3
Feral cat	0.8	0.9	0.1	0.7		
Initial sheep and vegetation density per ha	Seed (no./ha)	Sheep (no/ha)	Pasture (kg/ha)	Mature tree (no./ha)	Seedling (no./ha)	Sapling (no./ha)
	3,000	8	5,000	30	300	80

The size of the remnant is 50 ha and the farmer can put any proportion of the remnant under conservation. Source: Values on growth rate, interaction coefficients and carrying capacity are hypothetical. Values on initial population density of the animals, sheep and vegetation are collected from CSIRO.

1.2. Economic Component of the Model

Interventions affecting the life cycle of the species and carrying capacities of the site could be many. Retention of remnants (REM) that support the target species and their habitats is pre-requisite for their conservation. Management actions, such as fencing, large tree retention, stock exclusion, and revegetation or in-planting, are most commonly practiced by farmers in conservation programs in Australia. For our model, we have selected following types of management actions, which could be practiced on the patch (Figure 6) –

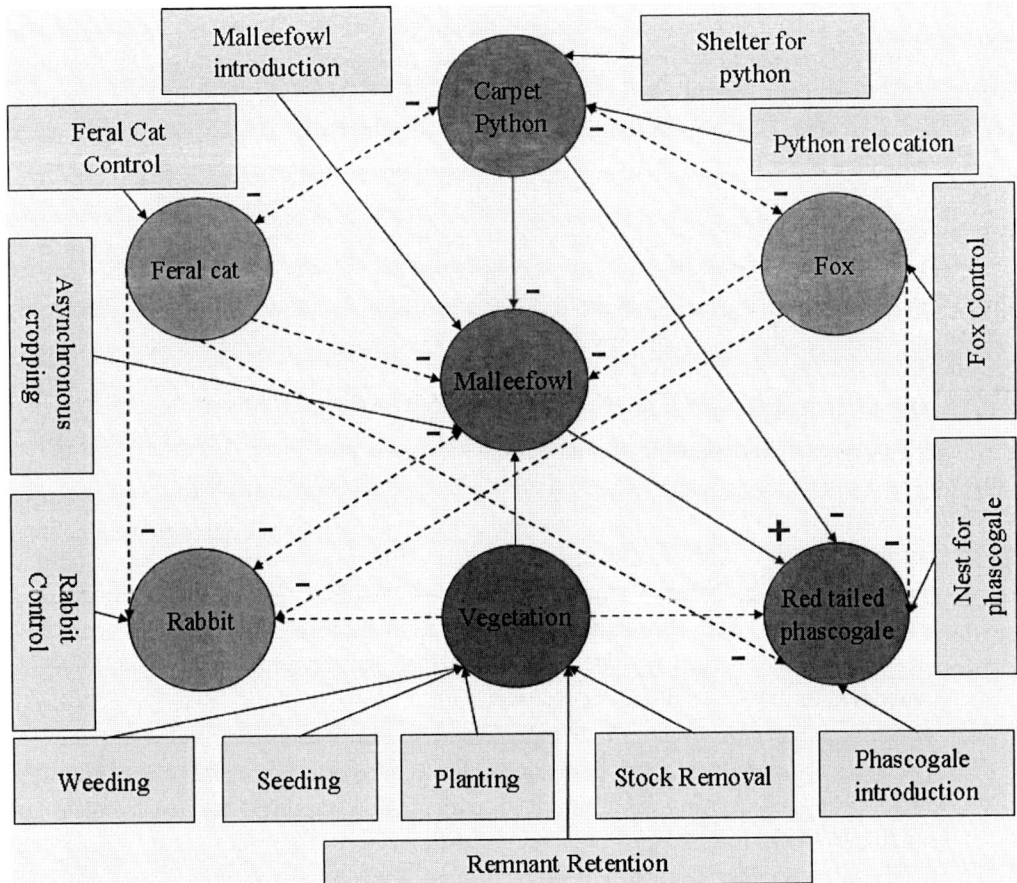

The (-) and (+) signs indicate negative and positive impacts respectively. The impact of prey populations on predator has been modelled through carrying capacity.

Figure 6. Interactions among the animals in the model.

- *Retention of remnants (REM):* Retention of remnants (REM) that support the target species and their habitats is pre-requisite for their conservation. The decision on the amount of land put under conservation scheme will depend on the opportunity cost of the remnant. It is equal to the monetary value of the alternative uses or activities the farmer is undertaking on the land. The farmer could use the remnant for grazing stock; collection of gravel, fence posts and firewood; and mining gravel for private use (Arnold, 2007). We have assumed that the size of square remnant is 50 ha and the farmer can decide to put any proportion of the remnant under conservation scheme. However, average per ha cost for remnant retention increases with the increase in amount of the remnant put under conservation.

- *Revegetation / in-planting with native species (REV)*: In-planting of the remnant with native species can improve the quality of remnant through provision of additional food and habitat. The farmers can practice two types of in-planting – seeding or planting. Direct seeding will increase the number of seeds available for germination in the next season and planting will increase the number of seedlings. It has been assumed that the farmer can sow maximum 2500 seeds per ha and plant 1200 seedlings per ha. The cost of seeding is much less but the survival probability is low (Dorrough et al., 2007), with 40% of the sowed seeds germinating. On the other hand, the cost of planting is high but survival probability is also high. It has been assumed that 70% of the planted seedlings will survive. Average per unit cost of revegetation reduces with the increase in area under conservation.

- *Weeding / removal of non-native species (WEED)*: Farmer could consider any vegetation (listed weeds or exotics) that compete with native vegetation as weeds. Weeds may be controlled by either mechanical or chemical means, but are usually sprayed with an herbicide (Arnold, 2007). Weeding would enhance survivability of the seedlings through reducing competition, though it may remove necessary cover and shelter for python and affect it negatively. Farmer can carry out weeding on any amount of area put under the conservation scheme. Average per unit cost of weeding reduces with the increase in area under conservation.

- *Asynchronous cropping (CROP)*: In some areas malleefowl often feed out from the edges of their habitat on fallen grain and green-pick. However, quite often crops are not grown every year around the remnants. Malleefowl would probably benefit if some accessible crops were grown each year. So keeping some cropland surrounding a remnant unharvested throughout the year could increase the carrying capacity of the remnant. It has been assumed that the farmer can maintain fallow land of a size equivalent to the area put under conservation.

- *Install nests for phascogale (N_PHASCOGALE)*: For phascogale, hollows in dead or live trees provide preferred den sites. However, hollow-bearing trees are often depleted by land clearing for urban development and agriculture and the formation of suitable hollows may take a long time. Artificial nest boxes can substitute for tree hollows, providing phascogale with nesting and roosting sites. The cost for purchasing and installing one nest box could be around $50. It has been assumed that the farmer can install a maximum of one nest per hectare. Average per unit cost of installation of artificial nest for phascogale reduces with the increase in area under conservation.

- *Stock exclusion (GRA):* Through stock exclusion the farmer can keep away the grazing stocks (such as sheep) and protect the remnant vegetation and enhance natural regeneration. The cost of stock removal will include material cost of removing the stock (e.g., fencing off the remnant) and the loss due to the removal of grazing stock from the remnant (Arnold, 2007).

- *Shelter for python (SHELT_PYTH)*: Hollow-bearing trees and logs, or large rock outcrops, plus thick litter or shrub cover, are crucial for carpet pythons. These are used as shelter sites, to avoid predators, to ambush prey, and to assist in thermoregulation (Shine and Fitzgerald, 1996). Such features provide habitat for prey items, particularly the herpetofauna utilised by juvenile pythons. Activities, which remove such shelter, threaten the survival of the python. So, in addition to the natural shelter, farmers may create artificial ground-layer habitats with piles of timber off cuts or discarded fence posts. These may provide shelter for python as well as for prey species for python (Lindenmayer et al., 2007). Farmer can provide shelter on any amount of area put under the conservation scheme. Average per unit cost of providing artificial shelter to pythons reduces with the increase in area under conservation.

- *Rabbit control (RAB)*: There are several options for rabbit control, such as baiting, shooting and trapping. Baiting is widely practiced in Australia. Several types of 1080 (sodium fluoroacetate) rabbit baits are available. 1080 is quickly broken down in the environment. In Western Australia, 'one-shot oat' technique is often used for rabbit baiting, where a small proportion ('one in a hundred' grains) of the oat bait contains sufficient 1080 to kill a rabbit. Free feeding is not required but the bait needs to be made available for an extended period and weather conditions need to be dry for 5-10 days. Bait is provided in a narrow trail at 6 kg/km (2–3 trails, 20 m apart), laid in a furrow or on the soil surface, or less frequently, in a broad (5m wide) scatter trail at 10 -12 kg/km. Both the bait and the rabbits should be left undisturbed for at least 10 days. To receive a lethal dose a rabbit would need to consume 11-32 gm of bait, whereas the average daily consumption of oat by rabbit is around 27 gm per rabbit (Twigg et al., 2002). Thus, one kilometre of trail would be able to kill around 240 rabbits in general, though the efficiency will depend on the density of rabbit. The material cost of baiting is around $10-12 per kilometre of trail. Average per unit cost of rabbit baiting reduces with the increase in area under conservation.

- *Fox control (FOX)*: In Western Australia, baiting with dried meat baits, which was originally developed for the control of dingoes (*Canis familiaris* dingo) (Thomson, 1986), are used for large-scale control of foxes. Studies indicate that on average 5 baits per sq km can effectively reduce fox numbers. At this density uptake of baits could be at a rate of 77-88% with an average of 80% and there is more than 95% probability that the fox would die within one day after consumption of bait (Thomson and Algar, 2000), unless the bait has somehow lost some toxicity. Thus, on average application of a bait would be able to kill 0.45 fox. The material cost of the bait would be $10 (at $2 per bait). It may be less if they make their own, or use a cheaper product at may be $1 per bait. It would take half an hour to drive around and choose places and lay the baits. Another 3 hours total spread over a 10-14 day period would require later keeping an eye on the baits and eventually retrieving uneaten ones. In total this would cost $6 per bait. Average per unit cost of fox control reduces with the increase in area under conservation.

- *Feral cat control (FER)*: Feral cat is also a major predator of malleefowl and phascogale. However, feral cat is an elusive animal to control. Techniques such as trapping and baiting for feral cat control are generally expensive, labour intensive and ineffective (Twyford et al., 2000). Shooting can be a humane method of removing feral cats when appropriately used (Mack et al., 2000). It has been estimated that on an average it would cost $10 to shoot a cat. Average per unit cost of feral cat control reduces with the increase in area under conservation.

- *Relocation of python*: If the farmer targets to conserve only prey species (malleefowl and / or phascogale) the python can be relocated from the project area to any other remnant, which is not under conservation contract to provide protection to the other two species. However, the farmer may need to get permission from the government. It has been assumed that due to these limitations the farmer would be able to relocate a maximum of 1 python from 2 hectares of remnant put under the conservation scheme.

- *Introduction of malleefowl*: Depending on the objective of the farmer the farmer could decide to enhance the existing population of malleefowl through introducing and relocating malleefowl in the project area. However, the farmer may need to get permission from the government. It has been assumed that due to these limitations the farmer would be able to introduce a maximum of 1 malleefowl per 2 hectares of remnant put under conservation scheme.

- *Introduction of phascogale*: Similarly, if the farmer is interested to maintain only phascogale population then they can be introduced into the remnant. However, as phascogale is strongly territorial there is a certain level up to which the farmer can introduce new phascogale. It has been assumed that due to these limitations the farmer would be able to introduce a maximum of 1 phascogale per 2 hectares of remnant put under the conservation scheme.

The type, unit, efficiency factor, per unit cost, and the minimum and maximum level of the interventions are presented in Table 5. Efficiency factor indicates the level of effectiveness of the intervention. For example, for planting, *ef* 0.7 indicates that out of 100 planted seedlings 70% will survive and add to the existing seedling stock in the remnant. Similarly, for fox baiting, *ef* 0.45 indicates that use of 100 baits will kill 45 foxes. For rabbit baiting, *ef* 240 indicates that use of 1 kilometre bait trail will kill 240 rabbits. In case of weeding, *ef* 0.3 means that the survival probability of seedlings will increase by 0.3 for each hectare of weeding. At the same time per ha weeding will reduce carrying capacity of the remnant for python by 0.01. On the other hand, creating shelters for python can increase the carrying capacity of the remnant by 2 per ha. Each hectare of asynchronous cropping around the remnant could increase the carrying capacity for malleefowl by 1.5 per hectare. Similarly, installation of one artificial nest could increase the carrying capacity for phascogale by 0.2. In case of introduction of malleefowl and phascogale, it has been assumed that there are 20% chances that the introduced animals will survive and add to the existing stock. It should be noted that efficiency of the animal control interventions depends on the density of the animals. Information on cost of the interventions has been collected from Department of Agriculture & Food, Western Australia and CSIRO.

Table 5. Costs and effectiveness of management options

Type	Unit	Efficiency factor (*ef*)	Per unit cost ($)	Minimum and Maximum level of interventions (per ha)
Remnant retention	Ha (Profit loss from alternative land use)	N/A	1,000 - 1000*4/sqrt(d*10,000)	max 50 ha
Revegetation [Direct seeding (*ds*)]	Number of seed	0.4	0.5	0 - 2500
Revegetation [Planting (*pl*)]	Number of seedling	0.7	2	0 – 1200
Stock exclusion (*sheep_rem*)	Number (Profit loss per sheep)	N/A	15	0 – Stock
Rabbit control [1080 baits trail (*r_bt*)]	Kilometre	240 Negative effect of baiting on fox: 0.01 Negative effect of baiting on python: 0.008	250 + 250*4/sqrt(d*10,000)	
Fox control [1080 baits (*fx_bt*)]	Number	0.45	6 + 6*4/sqrt(d*10,000)	
Feral cat control [Shooting (*fer_sh*)]	Number (cost to kill one cat)	N/A	10 + 10*4/sqrt(d*10,000)	
Asynchronous cropping for malleefowl [Asynch_crop (*crop*)]	Ha	1.5^	500	0 - 1
Shelter for python [Shelter_pyth (*shelter_pyth*)]	Ha	2^	300 + 4/sqrt(d*10,000)	0 - 1
Artificial nest for phascogale [Nest_phas (*N_phascogale*)]	Number	0.6^	50 + 50*4/sqrt(d*10,000)	0 - 3
Weeding	Ha	Effect of weeding on seedling: 0.3^^ Effect of weeding on sapling : 0.05^^ Negative effect of weeding on carrying capacity for python (ne_{weed}): 0.01	400 + 400*4/sqrt(d*10,000)	0 - 1
Relocation of python	Number	N/A	200	0 – 0.2
Introduction of malleefowl	Number	0.2	300	0 – 0.2
Introduction of phascogale	Number	0.2	250	0 – 0.2

*Here, d = area of the remnant put under conservation contract. ^ indicates the effect on carrying capacity for respective animals. ^^ indicates effect on survival rate of seedlings and saplings. Source: Values on the efficiency factors (*ef*) are hypothetical. Information on the cost, and minimum and maximum level of interventions has been collected from Department of Food and Agriculture, WA and CSIRO.

1.3. Interaction between Economic and Biological Components

The effect of the adopted interventions on the vegetation and wildlife in the patch can be divided into three types – on the vegetation, on the carrying capacity of the remnant and on the existing population. Interventions like seeding and planting can increase the density of the trees. Weeding can improve the survival probability of seedlings.

$$
\begin{aligned}
tr1_{m,t} &= tr4_{t-1} \times 100 - tr1(\text{mod})_t + ds_t \times ef_{ds} \\
tr2_{m,t} &= tr1_{t-1} \times 0.2 - tr2(\text{mod})_t + tr2_{t-1} \times (0.2 - tr2_{t-1} \times e^{-12.5} - tr3_{t-1} \times e^{-12} + weed_t \times ef_{weed}) \\
&\quad + pl_t \times ef_{pl} \\
tr3_{m,t} &= tr2_{t-1} \times (0.05 - tr2_{t-1} \times e^{-12}) + tr3_{t-1} \times (0.9 - tr3_{t-1} \times e^{-10} - tr4_{t-1} \times e^{-10}) - tr3(\text{mod})_t
\end{aligned} \tag{6}
$$

Here the subscript m indicates modified population size in time t. The modified population levels are used in the next period. ds, $weed$ and pl are variables for seeding, weeding and planting respectively. As it has been described earlier carrying capacity of the remnant for target species is influenced through interventions. Asynchronous cropping and installation of artificial nests can increase the carrying capacity for malleefowl and phascogale respectively. Carrying capacity for python could be reduced through weeding activities and enhanced through provision of artificial shelters.

$$
\begin{aligned}
K_{mallee} &= 0.00008 \times V_t + Crop_t \times ef_{crop} \\
K_{phas} &= tr2_t \times 0.0005 + tr3_t \times 0.0006 + tr4_t \times 0.0008 + N\_Phascogale_t \times ef_{N\_phascogale} \\
K_{pyth} &= k_{mallee\_pyth} \times h_{t-1}^{mallee} + k_{rabbit\_pyth} \times h_{t-1}^{rabbit} + k_{phas\_pyth} \times h_{t-1}^{phas} \\
&\quad + k_{cat\_pyth} \times h_{t-1}^{cat} + k_{fox\_pyth} \times h_{t-1}^{fox} + shelt\_pyth_t \times ef_{shelt\_pyth} - weed_t \times ne_{weed}
\end{aligned} \tag{7}
$$

Here, K_{mallee}, K_{phas}, K_{python} are carrying capacities for malleefowl, phascogale and python respectively. k_{mallee_pyth}, k_{rabbit_pyth}, k_{phas_pyth}, k_{cat_pyth} and k_{fox_pyth} are the interaction coefficients between python and malleefowl, rabbit, phascogale, feral cat and fox respectively. h^i is the population size of prey i. $Crop$, $N\_Phascogale$ and $shelt\_pyth$ are variables for asynchronous cropping, nest for phascogale and shelter for python respectively. The farmer can influence the population directly through some interventions. Through baiting the number of rabbits and foxes is reduced. Through shooting the number of feral cats is reduced. Through relocation she can reduce the number of pythons in the remnant. Some activities have negative effect on other animals. For example, rabbit baiting can negatively affect fox and python as both predate rabbit. On the other hand, malleefowl and phascogale can be introduced to increase their population size. She can also reduce grazing pressure on the remnant by removing stock from the remnant.

$$
sheep_{m,t} = sheep_t - sheep_{rem,t} \tag{8}
$$

$$h_{m,t}^{mallee} = h_t^{mallee} + \mathrm{mal\_int}_t \times ef_{\mathrm{mal\_int}}$$

$$h_{m,t}^{rabbit} = h_t^{rabbit} - r\_bt_t \times ef_{r\_bt}$$

$$h_{m,t}^{phas} = h_t^{phas} + \mathrm{phas\_int}_t \times ef_{\mathrm{phas\_int}} \qquad (9)$$

$$p_{m,t}^{fox} = p_t^{fox} - fx\_bt_t \times ef_{fx\_bt} - ne_{r\_bt\_fox} \times r\_bt_t \times p_{t-1}^{fox}$$

$$p_{m,t}^{cat} = p_t^{cat} - fer\_bt_t \times ef_{fer\_bt}$$

$$p_{m,t}^{python} = p_{t-1}^{python} - ne_{r\_bt\_pyth} \times r\_bt_t \times p_{t-1}^{python} - pyth\_rel_t \times ef_{pyth\_rel}$$

Here the subscript m indicates modified population size in time t. The modified population levels are used in next period. *mal_int, r_bt, phas_int, fx_bt, pyth_rel* and *fer_ct* are variables for malleefowl introduction, rabbit baiting, phascogale introduction, fox baiting, python relocation and feral cat control respectively. $ne_{r\text{-}bt\text{-}fox}$ and $ne_{r\text{-}bt\text{-}pyth}$ are variables for negative effect of rabbit baiting on fox and python respectively.

The model focuses solely on population interaction within the remnant and does not consider interactions with its buffer zone, and a single option for most of the interventions is considered. Often a combination of techniques can be more effective, for example, a combination of baiting, trapping, shooting and fencing was required to maintain areas free of feral cats and foxes (Risbey et al., 2000).

In: Ecological Restoration   ISBN 978-1-60741-013-3
Editors: George H. Pardue and Thomas K. Olvera   © 2009 Nova Science Publishers, Inc.

Chapter 2

RIVER CHANNEL RESTORATION FOR ECOLOGICAL IMPROVEMENT

Gemma L. Harvey and Nicholas J. Clifford
School of Geography, University of Nottingham, UK.

ABSTRACT

Recent decades have witnessed a growing appreciation among the academic and practitioner communities and the wider public of the importance of the ecological goods and services provided by freshwater systems, and the extent to which they have been degraded by human activity. In the case of river channels and their floodplains, a vast and diverse range of river restoration and rehabilitation schemes have been undertaken around the world in an attempt to remedy the misuse of water resources by humans in modern times. Many invaluable lessons have been learned through the implementation of such projects, in terms of both the practical realization of restoration efforts and the science which underpins them, which have broadened and deepened our understanding of the complexity and dynamism of river systems across a wide range of spatio-temporal scales. While there is cause for much optimisim, a range of scientific, technical, social, cultural and economic barriers persist and must be fully understood and addressed in order improve restoration success rates. This chapter explores the development of river restoration through the rise of environmentalism, discusses the scientific basis of restoration for ecological improvement, and examines some key issues associated with overall project success in terms of a range of ecological and socio-economic criteria. The final sections explore some of the key challenges and opportunities for further research in the river restoration science and the associated fields of the ecohydraulics and eco-hydromorphology.

INTRODUCTION: DEVELOPMENT OF THE RESTORATION PARADIGM IN RIVER SCIENCE

Human (Mis)Use of Rivers and Ecological Degradation

Riverine systems are the lifeblood of human civilizations, providing a range of goods and services which have been exploited by human populations for thousands of years (Newson, 1992; Buckley and Haddad, 2006; Alexander and Allan, 2007). Such goods and services include provision (e.g. food), support (e.g. waste processing, water supply) and cultural and social enrichment (e.g. through aesthetic and recreational qualities) for human populations (Giller, 2005). Thus, river systems are not simply associated with economic importance, but also with major environmental, social and cultural significance. The history of human use and management of river systems is long and complex, involving both direct 'planned' modifications the river channel, and 'unplanned' hydrological and geomorphological change induced indirectly through, for example, changes in land use within the floodplain and diffuse-source pollution (Hynes, 1970; Clifford, 2001). Perhaps the most obvious and large-scale motivations for human manipulation of river systems are associated with the provision of water supply, power supply, navigation, waste disposal and flood control; activities which, by the latter part of the twentieth century, had created a global network of rivers that had been impounded, regulated, diverted, straightened, confined, dredged and polluted (Downs and Gregory, 2004).

Until the middle of the twentieth century, the river management paradigm was one of controlling river systems and harnessing the power of their resources. Management took the form of hard engineering structures which suppressed natural dynamism and disconnected the longitudinal, lateral and vertical integration of river system components (Kondolf *et al.*, 1991; 2006). Dams and other control structures have disconnected the system longitudinally, reducing productivity and altering species composition in downstream reaches through adjustments to the water and sediment balance, channel morphology and chemical composition (Ward and Stanford, 1983; Petts *et al.*, 1985; Greenwood *et al.*, 1999; Graf, 2001; Kowaleswski *et al.*, 2000). Ecological communities have also suffered from the lateral disconnection of the river channel from the riparian corridor and floodplain as a result of flood defence and erosion control priorities. For instance, lateral disconnectivity through incision, dike and levee construction and reduced hydrological variability eliminates the changing conditions associated with the frequency, predictability and duration of flood pulses on which the maintenance of the structure, diversity and productivity of the biotic community depends (Junk *et al.*, 1989; Bayley, 1991). Simple maintenance of 'minimum instream flow requirements' for biota does not account for the provision of natural variability in the flow regime vital for natural aquatic ecosystem functioning (Bunn and Arthington, 2002). The diverse communities which permanently or temporarily inhabit the hyporheic zone at the groundwater-surface water interface are affected by the vertical disconnection of the system through hard engineering bed reinforcements, smothering of interstices with finer sediments and removal of obstacles such as woody debris which have important hydrogeomorphic influences (Boulton, 2007).

Conventional river management practices have direct and indirect impacts on the survival of aquatic organisms and the health of the aquatic ecosystem. Ongoing management practices, such as dredging of bed sediments or removal of aquatic vegetation to maintain flood capacity, can result in the direct removal of organisms from the channel (Aldridge, 2000), resulting in adjustments to community structure. Removal of vegetation and other flow obstructions also reduces the functionality of the physical structure of the river by removing food sources, oviposition sites, flow refugia and shelter from predators (Harper *et al.*, 1992). The physical effects of most channel modifications often involve an overall loss of physical diversity (Harvey and Wallerstein, in press), which is generally accepted to lead to a reduction in biological diversity (Smith *et al.*, 1995; Brookes and Shields, 1996) and resilience to disturbance (Townsend, 1989). Modifications to sediment transport capacity resulting from changes in the sediment balance, hydrological regime, or a combination of the two, can induce direct removal of organisms through scour, or habitat loss associated with erosion or smothering and burial (Argent and Flebbe, 1999; Soulsby *et al.*, 2001; Heywood and Walling, 2007). The competitive balance between species and predator-prey relationships may also be disrupted by direct modifications to the species composition through the introduction of non-native organisms (Manchester and Bullock, 2000). Problems in the UK, for instance, are associated with invasive plant species (*Fallopia japonica, impatiens glandulifer and Heracleum mantegazzianum;* Dawson and Holland, 1999) and non-native fish (e.g. *Pseudorasbora parva;* Britton *et al.,* 2007) and shellfish (*Pasifastacus leninsculus;* Crawford *et al.*, 2006; Guan and Wiles, 1997).

Setting the Stage for Restoration of River Ecosystems

The environmentally destructive effects of agriculture and urban developments on landscapes, resources and people were recognisable from the 17th century when the effects of deforestation on soil and water resources became apparent (Grove, 1992; 1995). Early concerns for the environment were philosophically as well as pragmatically informed, and Grove cites Alexander von Humbolt, the german geographer and explorer who was influenced by eastern philosophy, as a key founder of environmentalism. These ideas became more formalised into an environmental 'movement' in the mid- 19th century with the establishment of the national park system in the US and a focus on the importance of preserving 'wilderness' landscapes. Scholarly analyses of the destructive impacts of humans on the earth followed, exploring human-environmental relations and associated issues of resource stewardship, economic prosperity and ethical responsibilities (Marsh, 1864; Shaler, 1905). Issues of environmental protection and conservation were fully assimilated into the wider public conscience in the mid-20th century, supported by key milestone publications reporting on human impacts on the environment (e.g. Thomas, 1956; Carson, 1962); a growing the recognition of the importance of biodiversity for humanity (Stoll-Kleeman and O'Riordan, 2002); and an appreciation of the implications of physical, chemical and biological damage caused by anthropogenic disturbance for the provision of the ecosystem goods and services (Costanza *et al.* 1997; Baron *et al.*, 2002; Brismar, 2002). Habitat conversion for human usage was identified as the major threat to biodiversity at the global scale, and restoration ecology was offered as an opportunity to reverse the trend of biodiversity losses (Dobson *et al.*, 1997). More recently, a new and emerging paradigm of 'sustainability science' has been recognised, which examines the connections between the environment, development and societies through a new form of science: one that is regionally and locally-sensitive (Clark and Dickinson, 2003; Kates *et al.*, 2001) and which may have implications for widening the future scope of river restoration science (see later in this Chapter).

Changing attitudes towards the management of the natural environment more generally have been paralleled in river science, and the rapid growth of the environmental movement from the 1960s onwards facilitated the initiation of river restoration as both a management practice and a scientific discipline. The 1980s is often cited as a fundamental period in the development and establishment of river restoration as a discipline (Ormerod, 2004), marked by influential texts (e.g. Gore, 1988), the launch of academic journals with a focus on environmental conservation and restoration, and the reorganization of objectives and responsibilities in the water management arena (Gardiner, 1991). This period was characterised, in particular, by a transition from environmentally destructive 'hard' engineering management practices (e.g. channelization) to a greater emphasis on holistic and environmentally sensitive approaches (Park, 1981; Brookes, 1985; Gardiner, 1991); and a recognition of the need to understand the ways in which human use of rivers disrupts their natural functioning (Ward and Stanford, 1983), 'undo' past management (Brookes, 1987; 1990) and adopt a more ecologically sensitive approach to flood mitigation and energy production (LeClerc, 2002). Thus, rather than viewing morphological river channel adjustments as signs of instability which must be controlled, river restoration science now acknowledges that complexity and dynamism is a natural feature of fluvial systems (Gurnell and Petts, 1995; Petts *et al.*, 1995; Newson, 2002). The 1980s saw significant progress and an increased momentum in river restoration research and applications which has continued into the 21st century. However, there are important, but often overlooked, earlier examples, of restoration science with particular focus on channelization (e.g. Daniels (1960), Emerson (1971), Leopold (1969), Keller (1976) and Palmer (1976)), all of which helped to set the foundations for river restoration developments in subsequent decades.

Early endeavours in river restoration practice were generally opportunistic and *ad-hoc*, but have increased rapidly in popularity and scope from the late twentieth century (Skinner and Bruce-Burgess, 2005). Restoration projects have been implemented and reported across Europe, the Americas, Africa, Asia and Australasia (Boon *et al.*, 2000) with a wide network of demonstration projects developed to improve understanding and communicate findings (e.g. Nielsen, 1996; Holmes, 1998). Such developments have culminated in the establishment of bodies such as the UK River Restoration Centre (www.therrc.co.uk), the European Centre for River Restoration (www.errc.org), and the US National River Restoration Science Synthesis (NRRSS) working group (www.restoringrivers.org), reflecting a motivation for the standardisation of approaches and techniques, and widespread dissemination of knowledge and experiences. River restoration has now found a place at the forefront of applied river science (Wohl *et al.*, 2005) and represents a multibillion dollar industry, with widespread implementation of projects and a range of training frameworks and programmes (Rosgen, 1996; Brierley and Fryirs, 2005; Palmer *et al.*, 2007; Brooks and Lake, 2007).

The beginning of the twenty first century saw the introduction of international environmental legislation such as the EU Water Framework Directive (WFD) which recognises the need to protect and enhance the *ecological quality* of surface waters (The European Parliament and the Council of the European Union, 2000), compounding the efforts of previous decades of research and practice and driving restoration forward as a focal point for river research and management. Such direct recognition in the form of legislation enforces the societal values which have emerged over the course of the previous half century and beyond. Successful implementation of restoration efforts, however, requires a sound scientific understanding of river function; adequate monitoring, assessment, inventory and classification; and an integrated approach in terms of disciplines, policy and scale (Naiman *et al.*, 1992; Palmer *et al.*, 2000).

Motivations for river restoration are varied and often overlapping since many projects have multiple objectives (Wheaton, 2005). Alongside the enhancement of ecosystems and habitats, principal motivations may include the satisfaction of sediment management, erosion control, flood protection, aesthetic, recreational and educational goals (Wheaton, 2005; Wohl *et al.*, 2005; Bernhardt *et al.*, 2007; Darby and Sear, 2008). Ideally, all restoration efforts should have positive ecological benefits (Palmer *et al.*, 2005) and it is debateable as to whether those that do not should be referred to as 'restoration' projects at all (Gillilan *et al.*, 2005). The rest of this chapter focuses principally on the science behind ecological river restoration, but also explores the social, economic and cultural issues which underlie the overall success of restoration and must be incorporated into future research and practice.

THE SCIENTIFIC BASIS FOR RIVER CHANNEL HABITAT RESTORATION

A Spectrum of Ecological Improvement

The term 'restoration' has been used in the previous sections of this chapter to refer to the shift towards conservation goals in theoretical and practical river studies. The conservation paradigm, however, encompasses a combination of protection, preservation, restoration and sustainable use of river systems and their natural resources (Adams *et al.*, 2004). A distinction can therefore be made between the *enhancement* of a degraded system which refers to any form of measurable improvement; the *rehabilitation* of a system, which implies a partial restoration of previous form and structure; and *restoration*, which represents a complete return to the pre-disturbance state (Boon, 1992; Brookes, 1995a; Bradshaw, 1996; Figure 1). The complete restoration of a 'natural' state is generally considered impossible due to the difficulties of establishing the pre-disturbance condition and the continued human occupation of river basins (Ormerod, 2004; Downs and Thorne, 2000) and partial structural and functional return (rehabilitation) is usually adopted as the management goal. In some instances, where catchment hydrological and sediment systems have been dramatically altered by anthropogenic activity, the creation of a new ecosystem or any form of ecological enhancement may be a more practical option (Brieley and Fryirs, 2005; Figure 1). More recently, emphasis has been given to the concepts of assisted recovery or 'repair' of riverine systems (Brierley and Fryirs, 2008). For consistency with the title of this volume and previous influential texts, the term 'restoration' will be retained in this chapter, but is most frequently interpreted synonymously with the definition of rehabilitation provided above.

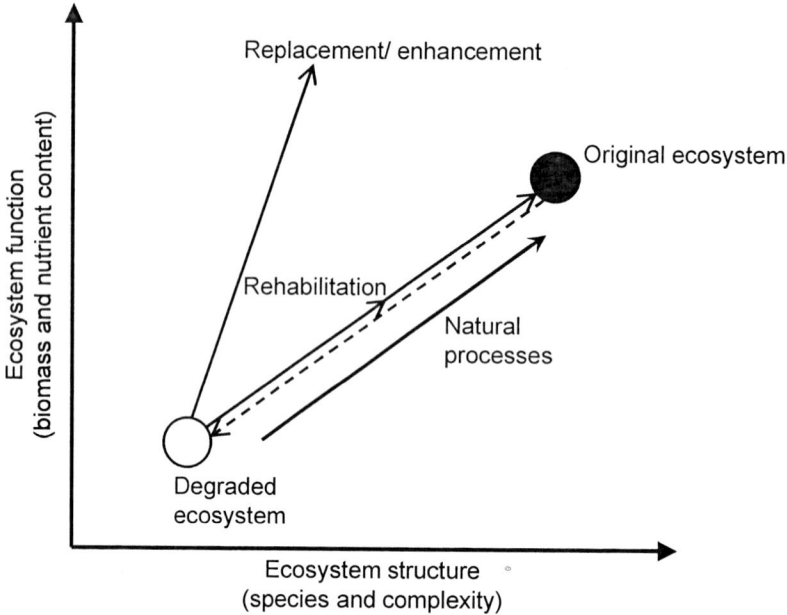

Source: Bradshaw (1999) with permission.

Figure 1. Options for the ecological remediation of river form and function.

The type of intervention required depends upon the severity of the modifications and the physical and ecological objectives (Boon, 1992; Brierley and Fryirs, 2005; Kondolf and Yang, 2008), although ultimately the action taken is dependent upon the availability of resources as well as the physical and ecological requirements. Pristine systems may simply require preservation of the current conditions, although such systems are extremely rare and most rivers have experienced some form of modification (Brookes, 1995a). Rivers with relatively low levels of habitat modification may require the limitation of catchment development or mitigation of the effects of modifications on habitats, while more heavily degraded sites will require some form of assisted recovery (Boon, 1992). Identification of geomorphic condition and potential recovery also provides a basis for prioritising restoration efforts at the catchment scale: successful conservation of 'intact' reaches may stimulate public and institutional support for the much more challenging (and costly) the restoration of severely degraded reaches (Brierley and Fryirs, 2005).

Urban river systems represent a unique case in restoration, particularly with relation to ecological goals due to their severe morphological and ecological degradation as a result of multiple environmental stressors acting at various scales: the 'urban stream syndrome' (Walsh et al., 2005). Physical degradation in urban rivers includes contamination of water and sediments, alteration of the hydrological regime and supply and transport of sediment, and changes to channel morphology (Gurnell et al., 2007). All of these factors influence the ecology of urban rivers, resulting in reduced biodiversity and a dominance of disturbance-tolerant species (Walsh et al., 2005). As a consequence urban streams require a different set of habitat quality assessment criteria and management strategies (Davenport *et al.*, 2004; Boitsidis *et al.*, 2006).

Regardless of the approach taken, intervention must be based on sound scientific knowledge. The following section explores the scientific basis for the ecological restoration of river channels through the use of ecohydraulics and the habitat level approach.

The Role of Ecohydraulics and Eco-Hydromorphology

This new paradigm of 'river restoration science' (Ormerod, 2004) necessitates interdisciplinary approaches to research problems, particularly involving close collaboration between scientists concerned with form and process in the physical channel environment (such as fluvial geomorphologists and hydrologists) and form and function in aquatic biology and ecology. Historically, these disciplines have followed separate research paths, both theoretically and empirically. Theoretical approaches from ecology tend to be extensive, deriving representative patterns and features from large 'populations', while geomorphological research is generally intensive in nature, exploring and attempting to explain the processes operating in individual, or small numbers of cases (Sayer, 1992; Richards, 1996). Empirically, ecologists have often considered the physical structure of the channel inferior to chemical properties in its affect on biota, while geomorphologists have focused much effort into characterising the physical structure of river channels (Leopold and Wolman, 1957; Schumm, 1985; Rosgen, 1994), but often with little regard to how this translates into habitat for aquatic biota. However, a need for true integration of disciplines has become increasingly important in the context of river management and conservation. Work towards the latter part of the 20th century addressed some of the interactions between hydrology, geomorphology and ecology, emphasising relationships between river form and dynamics and ecosystem structure and function. Key concepts include the river continuum concept which relates the influence of a longitudinal gradient in hydrogeomorphic conditions to the characteristics of biological communities (Vannote *et al.*, 1980); the role of physical disturbances for biodiversity and ecosystem resilience (Townsend, 1989; 1996); and the importance of a natural flow regime with 'flood pulses' for maintaining lateral connectivity with the floodplain and ecological integrity of river systems (Junk *et al.*, 1989; Poff *et al.*, 1997). Recognition of the importance of such links in maintaining healthy river ecosystems laid the foundations for interdisciplinary studies in the field of ecohydraulics.

The field of ecohydraulics has developed over the past decade, with the principal objective of restoring and protecting aquatic ecosystems through the physical enhancement of water courses, focusing on the abiotic factors contributing to habitats (LeClerc, 2002). Similarly, and as a result of the specification by the WFD that the achievement of 'good ecological status' for inland water bodies should be based on hydromorphological integrity, the term eco-hydromorphology has been adopted by geomorphologists to refer to similar interdisciplinary efforts (Clarke *et al.*, 2003; Newson and Large, 2006; Vaughan *et al.*, 2007). Whichever the chosen term of reference, the overriding goal is the integration of geomorphological, hydrological, hydraulic and ecological expertise to improve scientific understanding of interactions between physical and biological form and process and their implications for ecological status of river systems. Such work has, to date, been largely dominated by more deterministic and geocentric perspectives whereby the physical environment is considered to control the distribution of aquatic biota, but gradually the fuller integration of ecological and hydrogemorphological expertise is encouraging a wider acknowledgement of the complex interactions and feedbacks between the biotic and abiotic components of the instream environment (Corenblit *et al.*, 2007; Figure 2). For instance, aquatic plants and invertebrates demonstrate certain environmental preferences for flow and substrate conditions, but in turn alter the physical environment by changing particle size distributions, pore size and cohesion of the bed material; transferring matter and energy; and modifying local flow and sediment transport conditions (Newall, 1995; Kaenel *et al.*, 2000; Lopez and Garcia, 2001; Gurnell *et al.*, 2006; Naden *et al.*, 2006; Boulton, 2007; Corenblit *et al.*, 2007). Figure 2 illustrates some of the key interactions and feedbacks between aquatic vegetation and fluvial landform dynamics.

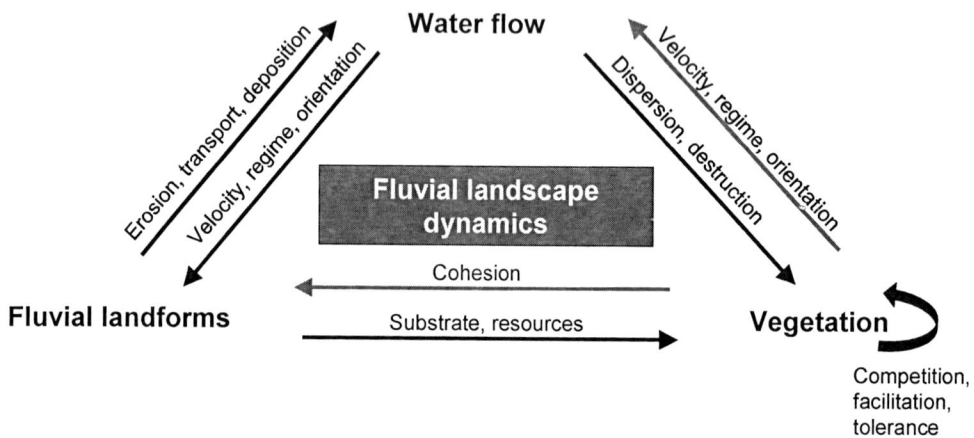

Source: Corenblit *et al.* (2007) with permission.

Figure 2. Interactions and feedbacks between hydrogeomorphic processes and vegetation dynamics in river systems.

Restoration and the Catchment Hierarchy

River catchments are complex ecological, hydrological, hydrochemical and geomorphological systems. The ecological organisation of streams is strongly related to physical variables such as temperature and channel hydraulics and a longitudinal gradient in physical conditions; and in sources, forms and processing of organic matter, is associated with changes in the structure and function of ecological communities (Vannote *et al.*, 1980). A combination of local spatial heterogeneity in physical variables and temporal heterogeneity in the form of disturbance regimes is superimposed onto the longitudinal continuum, creating a 'patchy' habitat structure (Southwood, 1977; Townsend, 1989). Furthermore, different components of the stream system are linked through a hierarchy of scales. Physical and biological processes operating at 'microscales' of several metres and over timescales of days or less have the most direct effect on the survival of individual biota (Biggs *et al.*, 2005). However, the wider 'mesoscale' reach morphology and associated hydraulics, variable over timescales of months and years, determines the community composition and is, in turn, controlled by the broader, 'macroscale' geological and climatic context of the catchment (Frissell *et al.*, 1986). Ideally, restoration should incorporate the whole river-floodplain system, focusing on restoring the natural hydrologic, geologic and riparian processes and connectivity (Roni *et al.*, 2002) and addressing system organization and response over different spatio-temporal scales (Poudevigne *et al.*, 2002). In practice, however, restoration has often been implemented in-stream and at the reach-scale as a result of land ownership issues, technical difficulties and economic constraints associated with whole-system restoration (Adams *et al.*, 2004).

The Habitat Level Approach

Within the catchment hierarchy, the 'mesoscale', which focuses on variation across the active channel width and along channel lengths that are small multiples of channel width, is often advocated as the most appropriate focus for instream habitat assessment and improvement programmes (Newson and Newson, 2000; Clarke *et al.*, 2003). However, mesoscale habitat level approaches must be nested within the catchment hierarchy, taking into account the catchment climatic, geological, geomorphological and ecological context for a particular reach of interest in order to ensure long-term sustainability and success (Skinner and Bruce-Burgess, 2005; Wharton and Gilvear, 2006).

The main attraction of the mesoscale is that it allows both scaling-up to the catchment and scaling-down to the microscale (Kershner and Snider, 1992), providing a 'fulcrum between scientific detail and universality' (Newson and Newson, 2000: 199). Strict ecological definitions suggest that the term 'habitat' refers to the abiotic surroundings of a particular *species* or *population*, while the term 'biotope' is used to refer to the physical environment of an ecological *community* (Udvardy, 1959). However, the term 'physical habitat' has been widely adopted among the research and practitioner communities as a means of describing the abiotic surroundings of instream biotic communities, as determined by the structure of the channel and the hydrological regime (Southwood, 1977; Maddock, 1999). Ultimately, the habitat level approach relies on the assumption that physical hydromorphological structure of the river channel determines biodiversity (Brookes and Shields, 1996), and while other constraints on ecosystem structure and function exist, both biotic (e.g. species composition and diversity; Chapin *et al.*, 1997) and abiotic (e.g. chemical water quality), it is generally accepted that mesoscale physical processes create a patchy mosaic of habitat units that perform different 'functions' for instream biota (Harper *et al.*, 1995; Nienhuis and Leuven, 2001). For instance, gravely riffles can provide spawning ground for fish (Garcia de Jalon, 1995), emergent macrophytes provide oviposition sites and passage to the water surface for emerging insects (Harper *et al.*, 1995) and low shear stress marginal channel areas can provide important refugia during spates (Lancaster and Hildrew, 1993).

Initial mesoscale approaches to habitat assessment and improvement focused on the (re)creation of favourable hydraulic conditions for specific target species using 'biological response models' (Mosely, 1982). A widely used model is PHABSIM, the one-dimensional 'Physical Habitat Simulation' model, developed by the US Fish and Wildlife Service as part of the Instream Flow Incremental Methodology (IFIM) decision-making framework for addressing instream flow issues (Bovee *et al.*, 1998). PHABSIM has been applied within the UK to support water resources decision-making (Spence and Hickley, 2000). The technique, however, identifies only the weighted usable area (WUA) of habitat for specific life stages of target species, and may therefore be undermined by community-level interactions such as competition and predation. This type of approach has been widely adopted by geomorphologists and hydrologists (Johnson *et al.*, 1995; Bockelmann *et al.*, 2004; Lane *et al.*, 2006), since they involve the operation of models which integrate biotic components without the requirement for specialist knowledge of ecological principles.

One-dimensional hydraulic habitat modelling approaches are, however, associated with a range of practical and theoretical limitations. From a practical perspective, results derived from models can vary significantly according to the overall methodological approach taken. For instance, habitat suitability predictions are sensitive to the choice of model (Paraseiwicz and Walker, 2007) and the appropriate use of Habitat Suitability Indices (HSIs): the use of more 'relaxed' habitat *utilization* rather than *preference* data can significantly alter results (Moir *et al.*, 2005). The spatial resolution of hydraulic data inputs also has a significant influence on model outputs, and the effects can vary significantly between habitat predictions for different life stages of the same species (Gard, 2005). In addition, the use of 1D hydraulic data does not account for three-dimensionality in hydraulic habitat, although some attempts have been made to address this through calculating weighted usable volumes rather than areas (Moulton *et al.*, 2007). Some of these limitations can be addressed through application of more complex 2D and 3D hydrodynamic models which can provide more detailed simulations of flow behaviour across a range of discharge conditions, although these are also associated with other practical uncertainties (Clifford *et al.*, 2008; Table 1).

Table 1. Application of common numerical flow models in river restoration identifying key issues associated with parameterisation and uncertainty.

Source: Clifford *et al.* (2008).

	Class	Applications	Parameters	Principal aspects of uncertainty
One-dimensional	HEC-RAS, ISIS backwater	Flood routing and channel conveyance Initial water surface elevation; velocity and depth between cross sections in PHABSIM	Downstream discharge; upstream water level; Manning's n; channel expansion and contraction coefficients, basin channel cross section morphology	Correct estimation of n and channel expansion and contraction coefficients; loss of information on cross-sectional flow distributional; no account of channel curvature.

Table 1. (Continued)

	Class	Applications	Parameters	Principal aspects of uncertainty
Two-dimensional	Telemac, RMA	Dynamic simulation of channel flow and floodplain inundation; cross-sectional patterns of habitat suitability	Boundary roughness (ks); flow information as above; turbulence closure model; choice of numerical solver and relaxation coefficients	Loss of information on depth-related flow properties; time-stepping in unsteady solutions; poor representation of secondary flow from channel discontinuities; meshing issues; representation of wetting and drying
Three dimensional	SSIIM, Fluent, CFX,	Detailed sub-reach modelling of habitat suitability	As above	Most of the above, plus: mesh-dependence in the vertical as well as cross-section; determination of the free water surface.

From a theoretical perspective, target species approaches can be undermined by ecological interactions such as competition and predation, geomorphological processes and water quality problems (Spence and Hickley, 2000; Clarke *et al.*, 2003): the fact that the required physical habitat conditions are present does not necessarily mean species are able to survive. Such approaches do not, therefore, provide a comprehensive assessment of overall ecosystem 'health' and functional value (Karr, 1999; Moss, 2000) and may even be counter-productive in river ecosystem protection and restoration (Moss, 2000). Furthermore, the physical habitat tolerances of aquatic biota are often very large and *relative* differences or *complexity* of physical habitat parameters and may represent a greater influence on organism distributions at patch scales as opposed to absolute values or target ranges (Lancaster *et al.*, 2006; Harvey and Clifford, 2008). Multi-species approaches focusing on target communities and multiple microhabitat parameters (e.g. Parasiewcz and Walker, 2007) offer an alternative to conventional one-dimensional models but require oversimplification of hydrological relationships and are largely focused on fish communities (Parasiewcz, 2007).

As a consequence of the issues outlined above, there is a growing trend towards a more holistic habitat level approach focused on communities of organisms rather than individual target species. This transition has led to the development of several mesoscale habitat concepts. Fluvial geomorphologists have generally taken a 'top-down' approach to mesoscale habitat characterisation, identifying units of channel morphology associated with different flow velocities, water depths and bed material sizes.

These features are commonly termed as 'physical biotopes' (Figure 3) and they refer to variations on the riffle-pool structure associated with the intermediate sized streams e.g. riffle, pool, run, glide, rapid, cascade.

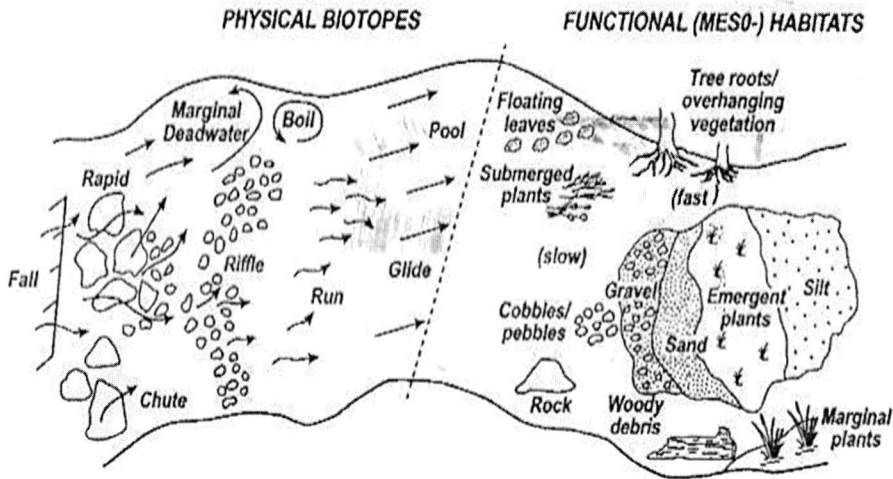

Figure 3. Physical biotopes and functional (meso-) habitat concepts for river habitat quality assessment and appraisal. Source: Newson and Newson (2000: 200) – Permission being sought.

Certain limitations to this approach exist, particularly associated with the discharge-dependency of relationships between channel morphology and hydraulics, variations in hydraulic characteristics between river types, and the research design and methodological approaches used (Clifford *et al.*, 2006). Early approaches to biotope characterisations (e.g. Jowett, 1993; Wadeson, 1994; Padmore, 1997) attempted to discretise elements of what is essentially a hydraulic continuum, and were therefore associated with limited success in discriminating between the hydraulics of different biotope units. More recent geomorphological investigations, however, support the assumption of *relative* variations in hydraulics associated with different geophysical mechanisms and suggest differing levels of complexity between units (Harvey and Clifford, 2008). Furthermore, a complementary ecologically-based 'bottom-up' approach to characterising 'functional' or 'meso-' habitats (Figure 3) associated with distinct assemblages of invertebrates (Harper *et al.*, 1992; Tickner *et al.*, 2000) provides vital ecological validation to the biotope concept (Kemp *et al.*, Harper *et al.*, 2000; Harvey *et al.*, 2008) and further supports its use in a river management context (Dyer and Thoms, 2006).

Research on a national habitat quality database in the UK (River Habitat Survey) identifies linkages between physical biotopes and functional habitats through a hierarchy of scales (Harvey *et al.*, 2008; Figure 4), improving the integrity of both concepts and providing a system for classifying and assessing river habitat status and potential in existing and restored channels. Such community-level approaches require further empirical testing through both large-scale field campaigns and smaller-scale process studies, but the biotope concept is rapidly gaining momentum within the research and practitioner community as an important focus for ecohydraulics research, an efficient means of assessing habitat quality, and a practical solution to requirements for large-scale resource cataloguing and appraisal.

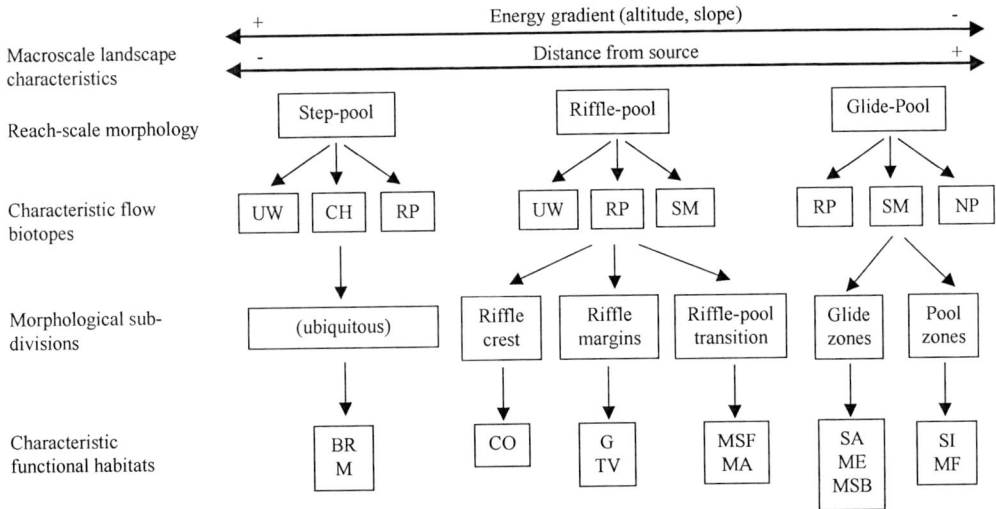

Source: Harvey *et al.* (2008: 649) with permission.

Figure 4. Hierarchical linkages between channel morphology, localised flow conditions and functional habitats, based on an analysis of data on semi-natural reaches derived from the UK River Habitat Survey database. (Biologically functional habitat units: BR = rocks, CO = cobbles, G = gravel, SA = sand, SI = silt, TV = trailing vegetation, ME = emergent macrophytes, MF = floating-leaved macrophytes, MSF = submerged fine-leaved macrophytes, MSB = submerged broad-leaved macrophytes, M = mosses, MA = macroalgae. Flow biotopes: NP = no perceptible flow, SM = smooth boundary turbulent, UP = upwelling, RP = rippled, UW = unbroken standing waves, BW = broken standing waves, CF = chaotic flow, CH = chute flow, FF = free fall.)

SUCCESS IN THE RESTORATION OF RIVERS FOR ECOLOGICAL PURPOSES

The potential benefits of river restoration are widely acknowledged. However, the continued scientific development and advance of 'adaptive' river restoration science relies heavily on learning from the success, and perhaps even more importantly, the failure of previous projects (Downs and Kondolf, 2002; Clarke *et al.*, 2003). This represents a challenge on several accounts since restoration efforts are poorly documented and monitored, or are inaccessible, and there is little agreement on what constitutes a 'successful' restoration attempt (Brooks and Lane, 2007; Giller, 2005). The following sections explore issues associated with achieving and measuring ecological success, followed by a discussion of the wider social, economic and cultural issues which must underpin the overall success of any restoration scheme.

Achieving Ecological Success

Restoration motivations are generally associated with environmental goals, at least in part, and success must be therefore measured according to whether ecological improvements have been made. Indeed, projects which do not achieve substantial ecological benefits may not be worthy of the term 'restoration' (Gillilan *et al.*, 2005). However, while ecological improvement is generally the motivation for projects, success is often measured on post-project appearance and public opinion (Bernhardt *et al.*, 2007). Palmer *et al.* (2005) suggest five criteria for assessing the ecological success of river restoration projects:

1. The articulation of a 'guiding image' that describes a dynamic, ecologically healthy endpoint based on historical information, reference sites, empirical geomorphological, hydrological and ecological knowledge, and river classification schemes.
2. The physicochemical and biological conditions of the river are measurably improved, which may include the recovery of populations and natural variability.
3. A more resilient and self-sustaining ecosystem is created, involving the restoration of natural river processes and preferably non-hard engineering methods.
4. Long term impacts are minimised and no lasting harm is done.
5. Ecological assessment is completed – both pre- and post- project implementation – and disseminated as widely as possible.

These criteria are generally supported by the academic and practitioner communities, with some suggested additions and amendments (Gillilan *et al.*, 2005; Jansson *et al.*, 2005) but represent significant challenges for river restoration scientists to overcome, some of which are acknowledged by the original authors. For instance, it has been suggested that river type - specific 'pristine' reference conditions are sought as a means of mitigating against uniform restorations which do not account for landscape, morphological and hydrological contexts (Muhar *et al.*, 1995), but finding adequate 'reference' reaches is inherently difficult as a result of significant changes to catchment land use and hydromorphology (Harvey and Wallerstein, in press). Furthermore it is generally considered inappropriate to attempt to emulate historical 'pristine' conditions in light of the continuing human occupation of basins (Rhoads and Herricks, 1996) and the long timescales of degradation and disconnection of river systems (Kondolf *et al.*, 2006): even those catchments and reaches with very limited anthropogenic disturbance and higher habitat quality may only realistically be regarded as semi-natural (Brookes, 1998; Graf, 2001). Whatever the guiding image is for a project, the ecological mechanisms by which the end state will be achieved should be identified and described to ensure that the success or failure of a project is explicitly understood (Jansson *et al.*, 2005), supported by pre-and post-project appraisals to develop transferable understanding (Downs and Kondolf, 2002). Even after an appropriate guiding image is devised, however, priorities may change during the project, and 'project hardening' (reduction in natural variability and adjustment as a result of the need to reduce risk, costs or repair efforts) may occur (Gillilan *et al.*, 2005).

Assessing the self-sustainability and resilience of a system is also inherently problematic since it suggests a requirement to compare the capacity of the system to recover from disturbance both before and after restoration work is undertaken (Jansson *et al.*, 2005). Furthermore, the establishment a self-sustaining system appears very difficult to achieve in practice: an estimated 60% of projects require on-going maintenance (Bernhardt *et al.*, 2007). Similarly, the assessment of both ecological improvements and 'harm' caused to the river system requires extensive post- project monitoring and appraisal plus comprehensive pre-project 'baseline' data (Skinner and Bruce-Burgess, 2005). A range of indicators may be used to assess the hydrogeomorphological, ecological and socio-economic outcomes of a project depending on the project objectives and scope (Woolsey *et al.*, 2007), but the outcomes of monitoring will be dependent upon timing of assessments and time lags in physical and ecological recovery. Long monitoring periods may be required. This often involves significant financial outlay at a stage of the project where budgets may be very limited.

Various components and timescales of Post Project Appraisal (PPA) have been identified, and Downs and Kondolf (2002) suggest that a 'full' PPA should ideally incorporate periodic or event-driven monitoring and historical and other secondary analyses over timescales of up to 10 years and beyond in order to facilitate adaptive management. Detailed monitoring and assessment of the geomorphological and ecological effects of restoration efforts is documented in the literature (e.g. Shields *et al.*, 1993; Gortz, 1998; Sear *et al.*, 1998; Pretty *et al.*, 2003; Millington and Sear, 2007; Klein *et al.*, 2007) but in general, comprehensive monitoring of restoration is lacking (Bernhardt *et al.*, 2007), particularly with respect to biological components (Alexander and Allan, 2007). Instead research and practioner communities have turned to rapid reconnaissance assessment methods as a more financially viable means of conducting PPA. Rapid appraisal technologies, such as the UK River Habitat Survey (RHS) methodology, offer a framework for conducting rapid (and therefore lower-cost) visual assessments of habitat quality, based on both geomorphological and ecological parameters, supported by simple quantitative measurements (Raven *et al.*, 1997). In the case of RHS, popular community-level habitat assessment concepts such as physical biotopes and functional habitat units are incorporated into the method. Such approaches require the simplification of a complex environment into rigid assessment categories, presenting problems for quantitative analysis, but do offer a useful tool for the development of river classification systems (e.g. Harvey *et al.*, 2008) which underpin river management and decision-making (O'Keefe and Uys, 2000). The RHS approach taken in the UK has led to the development of an extensive data resource of surveyed reaches which has been used extensively to support the scientific basis of river habitat assessment and restoration (Jeffers, 1998; Emery *et al.*, 2004; Davenport *et al.*, 2004; Boitsidis *et al.*, 2006; Harvey *et al.*, 2008; see Figure 4).

Socio-Economic and Cultural Dimensions to River Restoration

Ecological improvements generally provide a key motivation and focus for restoration efforts, and must be assessed appropriately. However, while restoration is often viewed as a scientific and technical problem, it has a fundamentally social basis since it reflects the societal decision to restore rivers, and requires stakeholder support for long term sustainable success (Connelly *et al.*, 2002; Wohl *et al.*, 2005). Thus, in addition to ecological goals, project success also involves cost-effectiveness, stakeholder acceptance and participation, recreational opportunities, the minimisation of any negative impacts and the advance of the science (Palmer *et al.* 2005; Woolsey *et al.*, 2007). Ideally a river restoration project will be successful in terms of all of these criteria, but in practice this is difficult to achieve.

Socio-economic constraints can include land use and ownership, government priorities, funding availability and willingness of financial sponsors, and value and belief systems (Wishart *et al.*, 2000; Buckley and Haddad, 2006; Alexander and Allan, 2007; Woolsey *et al.*, 2007). Such issues can represent major barriers to restoration efforts at the outset. For example, if land ownership is not conducive to restoration efforts, then ecological opportunities simply cannot be realised (Alexander and Allan, 2007; Bernhardt *et al.*, 2007). For this reason, many restoration projects to date have been opportunitistic, with sites chosen on the basis of landowner co-operation rather than ecological status. Administrative barriers exist as a result of the often fragmented management of river systems and funding allocations (Graf, 2001; Skinner and Bruce-Burgess, 2005). Some of these issues may be ameliorated in Europe by the WFD, which encourages management at the 'river basin district level' and, through legislative enforcement, requires member states to invest in measures to improve the ecological status of water bodies. However, different value systems in different parts of the world have an important influence on restoration priorities, and the type and extent of stakeholder support for restoration projects is partly a function of value and belief systems and associated philosophies of nature (Showers, 2000). Indeed, conservation may be considered an 'unaffordable luxury' by countries striving for economic development and alleviation of extreme poverty (Wishart *et al.*, 2000). At the local scale, variations in support among communities can be influenced by beliefs and environmental values as well as past involvement with environmental initiatives (Connelly *et al.*, 2000).

Community involvement in river restoration projects has been identified as a key factor in achieving overall project success (Bernhardt *et al.*, 2007), since river corridors represent an important local amenity associated with both organised and informal uses. Perceptions of river restoration practitioners and projects can be negative, and therefore involvement in the planning, implementation and post-project phases can help to improve understanding and foster unity and may help to remedy perceptions of self-interest associated with promotion of particular approaches, methods and personnel (Gillilan *et al.* 2005). Brierley and Fryirs (2005) stress the importance of prioritisation of projects in gaining public support for widespread investment in restoration schemes and suggest efforts should focus first on less degraded systems with higher potential for success, in order to gain community support for the more technically problematic and costly restoration of more severely degraded reaches in a catchment. Local residents attach importance to consultations and expect to be involved in the process (Tunstall *et al.*, 2000), and this is not only important at the design and implementation phases, of a project, but also subsequently in order to secure 'stewardship' of a restored reach following construction (Kondolf and Yang, 2008). Community support is therefore a fundamental element in the success of all restoration efforts, but is particularly vital in urban contexts where cultural and recreational values represent some of the principal ecosystem services provided by the river (Gurnell *et al.*, 2007).

Community involvement can, however, be problematic and it is vital that the expectations of different stakeholder groups are understood and addressed. Fundamentally, ecological goals may conflict directly with public perceptions of safety, cleanliness and even 'naturalness', reflecting a deep-rooted public preference for stable, grassy-banked meandering streams (Kondolf and Yang, 2008). Physico-chemical processes and biological functionality do not necessarily feature in public perceptions of the 'natural river' (Newson and Large, 2006) and indeed may be negatively perceived by communities. For instance, habitat features such as dense riparian vegetation and bank instabilities and irregularities may visually obscure the stream and affect access by the community, as well as leading to safety concerns associated with recreational use (Tunstall *et al.*, 2000; Kondolf and Wang, 2008). Similarly, the inclusion of wood in restoration projects as a means of restoring important ecological functions has been reported to provoke negative perceptions of disharmony, negligence and danger (Piegay *et al.*, 2005). Resistance to ecological outcomes of restoration projects may also be associated with changes in species composition, particularly with regard to recreational fishing, where reductions in non-native species may be viewed as a negative outcome. Thus, social and economic goals may constrain the amount ecological restoration possible for a particular project (Kondolf *et al.*, 2006) and must be adequately addressed and reconciled with ecological goals.

An emerging socio-cultural paradigm of restoration known as 'stream naturalisation' originates from the recognition that different interpretations of 'nature' must be acknowledged in the planning and implementation of river restoration projects. Rather than restoring pre-disturbance conditions, 'stream naturalisation' focuses on social visions of 'naturalness' and has been advocated as a more appropriate alternative to restoration and rehabilitation for heavily modified and/or resource-rich catchments where anthropogenic disturbance to the biophysical environment is likely to continue at similar rates at least in the medium-term (Rhoads *et al.*, 1999 Rhoads *et al.*, 2008). The approach is founded on the fact that the concept of nature is 'inescapably social' (Castree, 2001:3) and a 'natural' river is therefore a place-based social construct requiring any remedial action to incorporate social, as well as biophysical goals. While the concept is compatible with river restoration and rehabilitation approaches, it does not refer to any pre-disturbance condition and explicitly acknowledges, and attempts to incorporate, human utilization of water resources as a component of the natural environment (Rhoads and Herricks, 1996). The approach therefore incorporates a cultural dimension into river intervention, although this necessarily means that ecological goals form only one component of naturalisation objectives.

THE FUTURE

Advancing the Science

Notwithstanding significant scientific advances in the field of river restoration science, many avenues for further research exist. For instance, technological advances such as 2D and 3D hydrodynamic modelling can support more detailed simulations and analyses of hydraulic habitat provision at the reach scale of river systems which can be used to inform restoration design (Clifford *et al.*, 2002a; 2002b; Bocklemann *et al.*, 2004) and further development of multivariate and 'fuzzy' statistical approaches may contribute to the enhancement of the scientific basis for habitat assessment and appraisal (Lane *et al.*, 2006; Parasiewciz, 2007). Greater emphasis on process-based studies is encouraged (Vaughan *et al.*, 2007) and indeed will assist a move away from restoration of form, toward the restoration of self-sustaining processes (Wohl *et al.*, 2005).

However, true integration of ecology with geomorphology, hydrology and hydraulics, requires a focus on approaches which are respected by experts within all disciplines. The common use of habitat suitability criteria for target species in conjunction with hydraulic and hydrodynamic models is only of limited ecological value (see above) and may not represent the most robust way forward for the assessment, inventory, design and implementation of improved habitat quality for ecological goals. Physical biotopes and functional habitats may offer an alternative route, as a means of simplifying the complex interactions between geomorphology, hydrology and hydraulics in a way that holds ecological meaning. Similar directions are advocated from both ecologists and geomorphologists: the functional habitats approach is currently undergoing further development in order to incorporate species *traits*, rather than the species themselves, thus improving the transferability of the concept (D Harper, pers comm.), and geomorphological process studies emphasise the importance of *complexity*, *dynamism* and *relative* change in contributing to habitat integrity and resilience, as opposed to delineation of habitat features using absolute hydraulic ranges (Lancaster *et al.*, 2006; Harvey and Clifford, 2008). These trends are synthesised in Figure 5: the bottom left quadrant represents research focused on hydraulic parameterisation and delineation of hydraulic habitat units as a means of characterising habitats, coupled with a focus on the preservation of individual species or lifestages. A move towards 'higher order' parameters and the quantification of complexity, dynamism and resilience in the physical domain, and associated biologically 'functional' value in the ecological domain is represented in the top right quadrant. Research falling into bottom right and top left quadrants can also contribute to the scientific underpinnings of river restoration, through process-based experimental studies of habitat use by different species and/or lifestages, and explorations of habitat use by functional groups (rather than species) respectively. In the technocratic environment of numerical model-based approaches, however, 'softer' technologies may be resisted and a key challenge may lie in the reconciliation of quantitative model-based analyses with more holistic (and qualitative) assessments of habitat quality. In particular, work must focus on strengthening the understanding of links between geomorphology and physical biotopes, and between physical biotopes and ecological structure.

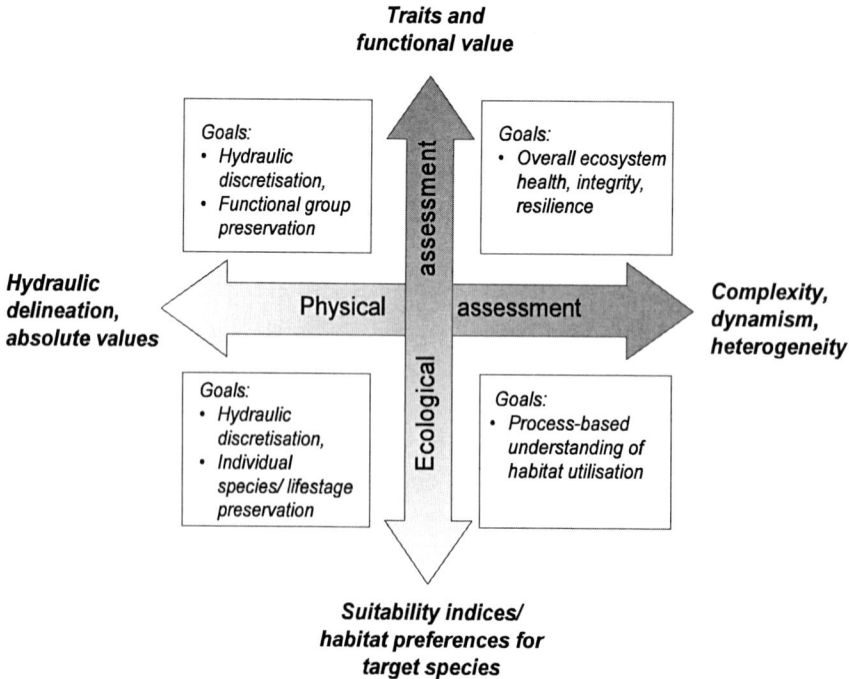

Figure 5. Directions in hydrogeomorphological and ecological assessment of river habitat.

Several authors stress the importance of a sound geomorphological understanding in successful, self-sustaining river restoration efforts (Brookes, 1995b; Soar and Thorne, 2001; Clarke *et al.*, 2003; Clifford, 2003; Newson, 2002; Newson *et al.*, 2006; Vaughan *et al.*, 2007). Often, geomorphological inputs to river restoration have taken the form of hydraulic and sedimentological design principles geared towards achieving geomorphic stability, and hence a physically self-sustaining system (Clifford and French, 1998; Shields *et al.*, 2003). Such inputs are vital to the success of restoration schemes, but the challenge for the future is to now integrate geomorphology more fully into the restoration of riverine ecosystems. Widespread acknowledgement of the valuable contributions from ecohydraulics and eco-hydromorphology provides a setting for this, and a range of geomorphological concepts and tools have been proposed (Newson, 2002). A particularly important direction is the development of a sound understanding of the ways in which geomorphological processes, and human-induced disruptions to them, affect ecological processes, rather than simply equating 'geodiversity' with biodiversity (Newson and Large, 2006).

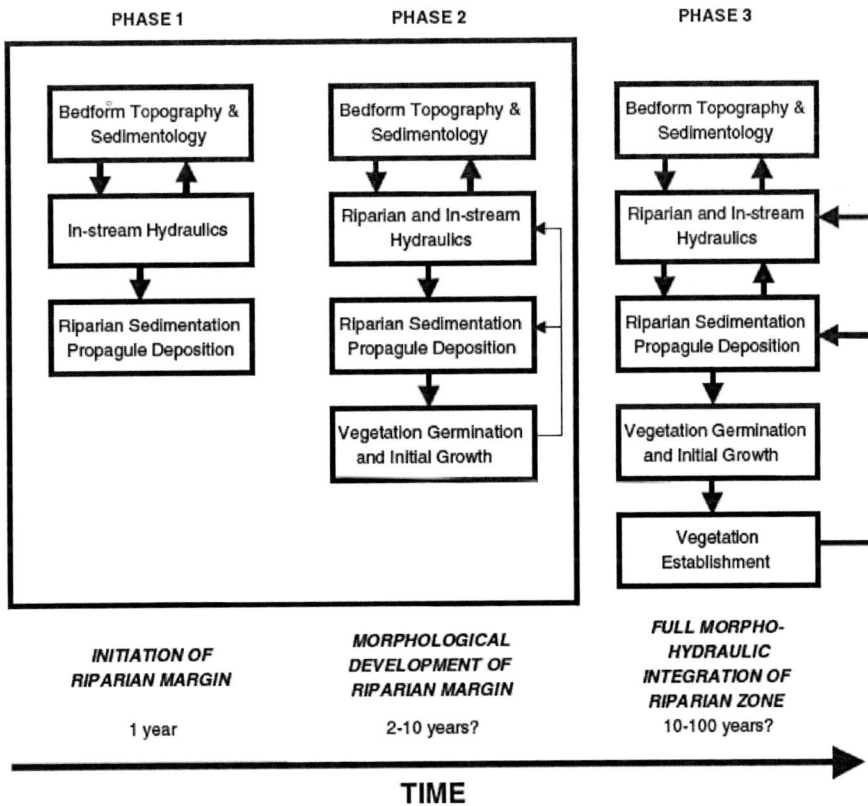

Source: Gurnell *et al.* (2006) with permission.

Figure 6. A conceptual model of interactions between hydraulics, sediment and propagule dynamics, and channel form in the development of newly created river margins. In the case of the River Cole, Birmingham, Phase 1 was achieved in less than a year following construction, Phase 2 after approximately 1 yea, and Phase 3 after approximately 2 years, suggesting that timescales may be much shorter than anticipated.

The river restoration community also must resist the propensity to over-manage, and pay attention to work which emphasises the ability of a river to 're-naturalise' itself following basic intervention. Recent experiences in a newly-cut reach of the River Cole, Birmingham, for instance, emphasise the importance of natural eco-hydromorphological processes, such as the transfer and establishment of riparian vegetation through hydrochory, which can create a heterogeneous instream and riparian environment over surprisingly short timescales (Gurnell *et al.*, 2006; Figure 6). Importantly, such an approach allows a river to adjust to a status which is 'natural' in the context of present-day conditions and is more likely to be self-regulating and sustainable in the long term (Van Rijen, 1998). Monitoring and dissemination of experiences is vital in communicating such information to the wider restoration community and further assessment of timescales associated with adjustment and recovery following restoration works is required to explore whether the rapid development of eco-hydromorphological processes at sites such as the River Cole is common across different contexts and typologies

Widening the Scope

It is widely acknowledged that the popular mesoscale-based approaches to river restoration must be supported by a sound understanding of the wider catchment geomorphological, hydrological and ecological context in order to be both successful and sustainable (Petts and Amoros, 1996). However, restoration efforts themselves are generally localised and focused at the reach-scale of small river channels as a result of the scientific, socio-economic and political complexities of conserving large river systems (Stanley and Boulton, 2000). As the field of river restoration continues to develop scientifically and gain further institutional (and financial) support, and as technological advances continue to increase data availability for large rivers and whole catchments, larger-scale projects will become more financially and practically feasible (Stanley and Boulton, 2000) particularly when supported by experiences from well-documented large-scale projects such as the Colorado and Kissimmee river restorations in the USA (Cohn, 2001; Whalen *et al.*, 2002). Coupled with international support for basin-wide sustainable management, which is particularly vital for catchments which traverse national boundaries, the scaling-up of river restorations from small, wadable streams to larger river systems which are currently under-represented in restoration efforts should become more widespread (Nienhuis and Leuven, 2001).

Concomitantly, further research is required into smaller-scale form-process interactions as a means of developing the scientific robustness of habitat level concepts such as the physical biotope. This can be achieved through the exploration of processes operating over small spatial- and temporal- scales within river systems, which can have the most direct consequences for the survival of individual organisms (Biggs *et al.*, 2005). For instance, turbulent properties, flow obstructions, secondary circulations and riparian vegetation and bank structure create hydraulic habitat niches at the microscale of river systems (Tsujimoto, 1996; Garcia de Jalon, 1995; Crowder and Diplas, 2000; Bisson *et al.*, 1981; Kellerhals and Miles, 1996). The analysis of high frequency flow properties within different river habitat units, for instance, suggests variations in mechanisms of turbulence generation, creating different hydraulic environments and levels of physical complexity which may play a role in influencing distributions of aquatic biota (Harvey and Clifford, 2008). Similarly, experimental work in process-geomorphology suggests variations between biotopes in terms of local sediment transport pathways associated with variations in hydraulics, with potential implications for the transfer of sediments, nutrients and pollutants at the biotope scale (Harvey and Clifford, in preparation).

As well as scientific and technological advancements, research into river 'naturalization' and the incorporation of cultural perceptions of nature, which may not be consistent with 'natural' geomorphic and ecological functionality, into restoration goals creates a new dimension to river restoration and thus a means of widening the scope across disciplines. Furthermore, the emerging wider 'sustainability science' paradigm offers a framework for widening the disciplinary scope of river restoration science and addressing environment and society interactions at different spatio-temporal scales.

Interaction, Communication and Interdisciplinary

It is widely acknowledged that there is a strong need for improved interaction both between academic disciplines involved in river restoration; and between the academy and practioners (Janauer, 2000; Giller, 2005). River restoration is necessarily a multidisciplinary science, bringing together expertise from fluvial geomorphology, hydrology, hydraulics and freshwater ecology. Each discipline is characterised by a unique cultural view and approach to the river system, resulting in some conflict of values, methodological approaches and scales of enquiry (Kondolf and Wang, 2008). Achieving true interdisciplinarity in river restoration science is perceived to be relatively rare, and requires a standardisation of terminology and monitoring protocols (for comparison and benchmarking) and the establishment of an improved interdisciplinary and inter-agency dialogue (Janauer, 2000; Alexander and Allan, 2007). At a general level, true interdisciplinarity can be difficult to achieve as a result of differing cultures, frames of reference, methods, disciplinary languages, and professional and institutional impediments (Brewer, 1999).

The communication of knowledge into legislation and policy can also be challenging (Palmer et al., 2000) and represents a considerable source of uncertainty in river restoration as a result of the (in)accessibility and meaning of language and the different expectations and philosophies of scientists and decision-makers (Graf, 2008). Thus, in the current legislative environment, improved communication is needed between scientists and decision-makers both for responding to legislative demands such as the WFD, and in the development of policies based on robust science. This, however, requires considerable investment of time and energy into developing mutually beneficial relationships (Rhoads et al., 1999). Similarly, it is vital that restoration goals are communicated appropriately to communities and stakeholders. For instance, local residents may show greater support for broader environmental goals than specific implementation actions and their associated technicalities (Connelly et al., 2002) and Wohl et al. (2005) emphasise the role of river scientists as 'public educators' with a responsibility for communicating scientific information and a capacity to influence societal values. It is acknowledged that social successes are vital to the future relevance and vitality of river restoration (Wohl et al., 2005), and therefore any real (or perceived) conflicts between ecological and human needs must be reconciled through partnerships and shared visions between scientists and other stakeholders (Poff et al., 2003)

A key challenge, and opportunity, for improved interdisciplinarity and communication, is associated with the reconciliation of ecological restoration goals with other human demands, such as the management of erosion and increasing flood risk, through multi-functional river management (Gilvear, 1999; Downs and Thorne, 1998; 2000). The need to manage increasing risks of flooding associated with climatic and socio-economic change is now acknowledged explicitly in international legislation (e.g. The European Parliament and the Council of the European Union (2007)) and represents an important statutory obligation for management bodies and decision-makers. Such management demands can represent a direct source of conflict with ecological restoration goals and require a sound scientific understanding of the interaction between flood defence infrastructure and aquatic habitat (Harvey and Wallerstein, in press). However, synergisms and significant potential for successful integration of management needs also exists (Nienhuis and Leuven, 2001). Indeed, the WFD incorporates the mitigation of flood risk through sustainable management practices (The European Parliament and the Council of the European Union, 2000) and there has been increasing recognition of the potential for utilising and re-establishing natural flood storage in wetlands and washlands which can have ecological benefits through reconnecting the channel to its floodplain and restoring natural water and sediment transfer processes (Wharton and Gilvear, 2006). Such efforts also support a move away from the species preservation approach towards the rehabilitation of the 'functional value' of the river system as described above. Furthermore, government budgets for flood defence works are necessarily extensive, and therefore restoration projects which can incorporate both ecological and flood defence benefits have the potential to unleash considerable funds which may not be available to ecological efforts alone (Adams *et al.*, 2004).

CONCLUSION

Recent decades have witnessed a growing appreciation among academic researchers, environmental managers and decision-makers, and the wider public, of the importance of the ecological goods and services provided by freshwater ecosystems, and the extent to which they have been degraded by human activity. As a result, a diverse range of river restoration schemes have been undertaken around the world in an attempt to remedy the longstanding human misuse of water resources. Scientific advances in the river restoration science research arena have broadened and deepened our understanding of the complexity and dynamism of river systems across a wide range of spatio-temporal scales. Despite such developments, however, river restoration success rates remain low as a result of both scientific and technical deficiencies, and a range of social, cultural, political and economic barriers.

This chapter has explored the initiation and development of river restoration for the purposes of ecological improvement; the scientific basis of river channel restoration for ecological goals; and some of the key issues associated with achieving ecological, social, cultural and economic 'success'. The final sections identify some of the key challenges and opportunities facing academics and practitioners, providing some focal points for future research.

From its roots in the environmental movement of the mid-20th century, river restoration is developing into a robust science and a multi-billion dollar management activity. Although popular and rapidly developing, however, it is still a relatively 'young' science. Both scientific and practical refinement of concepts, tools and techniques is required in order to ensure that restoration activities are successful in both the short- and longer-term; in terms of achieving ecological and socio-economic goals; and in terms of advancing the science of river channel restoration and the associated disciplines of ecohydraulics and eco-hydromorphology.

REFERENCES

Adams, W. M., Perrow, M. R. and Carpenter, A. (2004) Conservatives and champions: river managers and the river restoration discourse in the United Kingdom. *Environment and Planning* A, 36: 1929-1942.

Alexander, G. G. and Allan, J. D. (2007) Ecological success in stream restoration: case studies from the Midwestern United States. *Environmental Management* 40: 245–255

Argent DG, Flebbe PA. 2007. Fine sediment effects on brook trout eggs in laboratory streams. *Fisheries Research* 39: 253-262.

Baron, J. S., LeRoy Poff, N., Angermeier, P. L, Dahma, C. N., Cleick, P. H., Hairston Jr., N. G., Jackson, R. B., Johnston, C. A. (2002) Meeting ecological and societal needs for freshwater. *Ecological Applications* 12 (5): 1247-1260.

Bernhardt, E. S., Sudduth, E. B., Palmer, M. A., Allan, J. D., Meyer, J. L., Alexander, G., Follastad-Shah, J., Hassett, B., Jenkinson, R., Lave, R., Rumps, J. and Pagano, L. (2007) Restoring rivers one reach at a time: results from a survey of U.S. river restoration practitioners. *Restoration Ecology* 15 (3): 482-493.

Biggs, B. J. F., Nikora, V. I., Snelder, T. H. (2005) Linking scales of flow variability to lotic ecosystem structure and function. *River Research and Applications* 21: 283–298.

Bisson, P. A., Nielson, J. L., Palson, R. A. and Grove, L. E. (1981) A system of naming habitat in small streams, with examples of habitat utilization by salmonids during low streamflow. Acquisition and Utilization of Aquatic Habitat Inventory Information: Proceedings of A Symposium, American Fisheries Society Western Division, Bethesda, Maryland.

Bocklemann, B. N., Fenrich, E. K., Lin, B. and Falconer, R. A. (2004) Development of an ecohydraulics model for stream and river restoration. *Ecological Engineering*, 22: 227-235.

Boitsidis, A.J., Gurnell, A.M., Scott, M., Petts, G.E., Armitage, P. A (2006) Decision support system for identifying the habitat quality and rehabilitation potential of urban rivers. *Water and Environment Journal*, 20: 1-11.

Boon, P. J. (1992) Essential elements in the case for river conservation. In *River Conservation and Management*, Boon, P. J., Calow, P. and Petts, G. (eds.), John Wiley and Sons: Chichester. P. 11-33.

Boon, P. J. (1998) River restoration in five dimensions. *Aquatic Conservation: Marine and Freshwater Ecosystems*, 8: 257-264.

Boon, P. J., Davies, B. R. and Petts, G. E. (eds.) (2000) *Global Perspectives in River Conservation*. John Wiley and Sons: Chichester.

Boulton, A. J. (2007) Hyporheic rehabilitation in rivers: restoring vertical connectivity. *Freshwater Biology* 52:632-650.

Bovee, K. D., Lamb, B. L., Bartholow, J. M., Stalnaker, C. B., Taylor, J. and Henriksen, J. (1998) Stream Habitat Analysis Using the Instream Flow Incremental Methodology Information and Technology Report, Fort Collins, CO: U.S. Geological Survey-BRD.

Bradshaw, A. D. (1996) Underlying principles of restoration. *Canadian Journal of Fisheries and Aquatic Sciences* 53: 3-9.

Brewer, G. D. (1999) The challenges of interdisciplinarity. *Policy Sciences* 32: 327-337.

Brierley, G. J. and Fryirs, K. A. (2005) *Geomorphology and River Management: Applications of the River Styles Framework*. Blackwell Publishing.

Brierley, G. J. and Fryirs, K. A. (2008) River Futures: An Integrative Scientific Approach to River Repair. Island Press: Washington.

Brismar, A. (2002) River systems as providers of goods and services: A basis for comparing desired and undesired effects of large dam projects. *Environmental Management* 29 (5): 698-609.

Brooks, S. S. and Lake, P. S. (2007) River restoration in Victoria, Australia: change is in the wind, and none too soon. *Restoration Ecology* 15 (3): 584-591.

Brookes, A M. (1985) River channelisation: traditional engineering methods, physical consequences and alternative practices. *Progress in Physical Geography* 9, 44-73.

Brooks, A. M. (1987) Restoring the sinuosity of artificially straightened stream channels. *Environmental Geology and Water Sciences* 10, 33-41.

Brookes, A. M. 1990 Restoration and enhancement of engineered river channels: some European examples. *Regulated Rivers: Research and Management*, 5, 45-56.

Brookes, A. (1995a) River restoration theory and practice. In *Changing River Channels*, Gurnell, A. M. and Petts, G. E. (eds), John Wiley and Sons Ltd: Chichester. P. 369-388.

Brookes, A. (1995b) Challenges and objectives for geomorphology in UK river management. *Earth Surface Processes and Landforms*, 20(7): 593-610.

Socially Strategic Ecological Restoration: A

Buckley, M. and Haddad, B. M. (2006) Game-theoretic analysis shortened: socially strategic restoration. *Environmental Management* 38 (1): 48-61.

Bunn, S. E. and Arthington, A. H. (2002) Basic principles and ecological consequences of altered flow regimes for aquatic biodiversity. *Environmental Management* 50: 492-507.

Carson, R. (1962) *Silent Spring*. Houghton Mifflin: Boston.

Castree, N. (2001) Socializing Nature: Theory, Practice, and Politics. In *Socializing Nature: Theory, Practice, and Politics*, Castree, N. and Braun, B. (eds.), Blackwell Publishing.

Chapin III, F. S., Walker, B. H., Hobbs, R. J., Hopper, D. U., Lawton, J. H., Sala, O. E. and Tilman, D. (1997) Biotic Control over the Functioning of Ecosystems. *Science* 277 (5325): 500 – 504

Clifford, N. J. and French, J. R. (1998) Restoration of channel physical environment in smaller, moderate gradient rivers: geomorphological bases for design criteria. In *United Kingdom Floodplains*, Bailey, R. G., Jose, P. V. and Sherwood, B. R. (eds), Westbury Publishing. P. 63-82.

Clifford, N. J. (2001) Conservation and the River Channel Environment. In *Habitat Conservation: Managing the Physical Environment*, Warren, A. and French, J. R. (eds), John Wiley and Sons Ltd: Chichester. P. 67-104.

Clifford, N. J., Soar, P. J., Emery, J. C., Gurnell, A. M. and Petts, G. E. (2002a) Sustaining water-related ecosystems - the role of in-stream bedform design in river channel rehabilitation. In: Friend 2002 - Regional hydrology: bridging the gap between research and practice (Proceedings of the Fourth International FRIEND Conference): 407-413.

Clifford, N. J., Soar, P. J., Petts, G. E., Gurnell, A. M. and Emery, J. C. (2002b) Numerical flow modelling for eco-hydraulic and river rehabilitation applications: a case study of the River Cole, Birmingham, UK. In: Bousmar, D. and Zech, Y. (Eds.) River Flow 2002, Lisse, Swets & Zeitlinger/Balkema: 1195- 1204.

Clifford, N. J. (2003) River restoration: an overview of approaches from a habitat and geomorphological perspective. Proceedings of the Institute of Fisheries Management 34th Annual Study Course, September 2003: 79-97.

Clifford, N. J., Harmar, O. P, Harvey, G., Petts, G. E. (2006) Physical habitat, eco-hydraulics and river design: a review and re-evaluation of some popular concepts and methods. *Aquatic Conservation: Marine and Freshwater Ecosystems* 16 (4): 389-408.

Clifford, N. J., Acreman, M. C. and Booker, D. J. (2008) Hydrological and hydraulic aspects of river restoration uncertainty for ecological purposes. In *River Restoration: Managing the Uncertainty in Restoring Physical Habitat*, Darby, S. and Sear, D. (eds). John Wiley and Sons Ltd: Chichester.

Clark, W. and Dickson, N. M. (2003) Sustainability science: the emerging research program. *Proceedings of the National Academy of Sciences,* 100, 8059-8061.

Clarke, S. J., Bruce-Burgess, L. and Wharton, G. (2003) Linking form and function: towards an eco-hydromorphic approach to sustainable river restoration. *Aquatic Conservation: Marine and Freshwater Ecosystems*, 13: 439-450.

Cohn, J. P. (2001) Resurrecting the dammed: a look at the Colorado river restoration. *Bioscience* 51 (12); 998-1003.

Connelly, N. A., Knuth, B. A. and Kay, D. L. (2002) Public support for ecosystem restoration in the Hudson River Valley, USA. *Environmental Management* 29 (4): 467-476.

Corenblit, D., Tabacchi, E., Steiger, J. and Gurnell, A. M. (2007) Reciprocal interactions and adjustments between fluvial landforms and vegetation dynamics in river corridors: A review of complementary approaches. *Earth-Science Reviews* 84: 56-86.

Costanza, R., d'Arge, R., de Groott, R., Farber, S., Grasso, M., Hannon, B., Limburg, K., Naeem, S., O'Neill, R. V., Paruelo, J., Raskin, R. G., Sutton, P. and van den Belt, M. (1997) The value of the world's ecosystem services and natural capital. *Nature* 387: 253-260.

Crawford, L., Yeomans, W. E. and Adams, C. E. (2006) The impact of introduced signal crayfish Pacifastacus leniusculus on stream invertebrate communities. *Aquatic Conservation: Marine and Freshwater Ecosystems* 16: 611-621.

Crowder, D. W. and Diplas, P. (2000) Using two-dimensional hydrodynamic models at scales of ecological importance. *Journal of Hydrology* 230: 172-191.

Daniels, R B. 1960 Entrenchment of the Willow Drainage Ditch. *American Journal of Science* 258, 161-76.

Darby, S. and Sear, D. (eds) (2008) *River Restoration: Managing the Uncertainty in Restoring Physical Habitat*. John Wiley and Sons Ltd: Chichester.

Davenport AJ, Gurnell AM. Armitage PD. 2004. Habitat Survey and Classification of Urban Rivers. *River Research and Applications* 20: 687-704.

Dawson, F. H. and Holland, F. (1999) The distribution in bankside habitats of three alien invasive plants in the U. K. in relation to the development of control strategies. *Hydrobiologia* 415: 193-201.

Dobson, A. P., Bradshaw, A. D. and Baker, A. J. M. (1997) Hopes for the Future: Restoration Ecology and Conservation Biology, *Science* 277 (5325): 515 - 522

Downs, P. W. and Kondolf, G. M. (2002) Post-project appraisals in adaptive management of river channel restoration. *Environmental Management*, 29: 477-496.

Downs, P. W. and Thorne, C. R. (1998) Design principles and suitability testing for rehabilitation in a flood defence channel: the River Idle, Nottinghamshire, UK. *Aquatic Conservation: Marine and Freshwater Ecosystems* 8: 17-38.

Downs, P. W. and Thorne, C. R. (2000) Rehabilitation of a lowland river: reconciling flood defence with habitat diversity and geomorphological sustainability. *Journal of Environmental Management* 58: 249-268.

Downs, P. W. and Gregory, K. J. (2004) *River Channel Management: Towards Sustainable Catchment Hydrosystems*. Arnold: London.

Dyer F. J. and Thoms M. C. (2006) Managing river flows for hydraulic diversity: an example of an upland regulated gravel bed river. *River Research and Applications* 22: 257-267.

Emerson, J. W. (1971) Channelisation, a case study. *Science* , 173, 325-6.

Emery, J. C., Gurnell, A. M., Clifford, N. J., Petts, G. E., Morrissey, I. P. and Soar, P. J. (2003) Classifying the hydraulic performance of riffle-pool bedforms for habitat assessment and river rehabilitation design. *River Research and Applications* 19: 533-549.

Frissell, C. A., Liss, W. J., Warren, C. E. and Hurley, M. D. (1986) A hierarchical framework for stream habitat classification: viewing streams in a watershed context. *Environmental Management* 10: 199-214.

Garcia de Jalon, D. (1995) Management of physical habitat for fish stocks. In: D. M. Harper and A. J. D. Ferguson (Ed.s) *The Ecological Basis for River Management*. Chichester, John Wiley and Sons: 363-374.

Gardiner, J. L. (1991) *River Projects and Conservation: A Manual for Holistic Appraisal*. John Wiley & Sons Ltd: Chichester.

Giller, P. S. (2005) River restoration: seeking ecological standards. *Journal of Applied Ecology* 42: 201-207.

Gillilan, S., Boyd, K., Hoitsma, T. and Kauffman, M. (2005) Challenges in developing and implementing ecological standards for geomorphic river restoration projects: a practioners response to Palmer *et al.* (2005). *Journal of Applied Ecology* 42: 223-227.

Gore, J. A. (Ed.) (1985) *The Restoration of Rivers and Streams: Theories and Experience*. Butterworth: Ann Arbor.

Gortz, P. (1998) Effects of stream restoration on the macroinvertebrate community in the River Esrom, Denmark. *Aquatic Conservation: Marine and Freshwater Ecosystems* 8: 115-130.

Graf, W. L. (2001) Damage control: restoring the physical integrity of America's rivers. *Annals of the Association of American Geographers*, 91(1): 1-27.

Greenwood, M. T., Bickerton, M. A., Gurnell, A. M. and Petts, G. E. (1999) Channel changes and invertebrate faunas below Nant-Y-Moch Dam, River Rheidol, Wales, UK: 35 years on. *Regulated Rivers Research and Management* 15: 99-112.

Grove, R. H. (1992) Origins of Western Environmentalism. *Scientific American* 267: 22-27.

Grove, R. H. (1995) Green Imperialism: Colonial Expansion, Tropical Island Edens and the Origins of Environmentalism, 1600-1860. Cambridge University Press: Cambridge and New York.

Guan, R. Z. and Wiles, P. R. (1996) Ecological impact of introduced crayfish on benthic fishes in a British lowland river. *Conservation Biology* 11 (3): 641–647

Gurnell, A. M. and Petts, G. E. (eds) (1995) *Changing River Channels.* John Wiley and Sons Ltd: Chichester.

Gurnell, A. M., Morrissey, I. P., Boitsidis, A. J., Bark, T., Clifford, N. J., Petts, G. E. and Thompson, K. (2006) Initial adjustments within a new river channel: interactions between fluvial processes, colonizing vegetation, and bank profile development. *Environmental Management*, 38: 580-596.

Gurnell, A. M., Lee, M. and Souch, C. (2007) Urban rivers: hydrology, geomorphology, ecology and opportunities for change. *Geography Compass* 1/5: 1118-1137.

Harper, D. M., Smith, C. D. and Barham, P. J. (1992) Habitats as the building blocks for river conservation assessment. In *River Conservation and Management*, Boon, P. J., Calow, P. and Petts, G. E. (eds.), John Wiley and Sons Ltd: Chichester. P. 311-319.

Harper, D. M., Kemp, J. L., Vogel, B. and Newson, C. L. (2000) Towards an assessment of 'ecological integrity' in running waters of the United Kingdom. *Hydrobiologia* 422/423: 133-142.

Harvey, G., Clifford, N. J. and Gurnell, A. M. (2008) Towards an ecologically meaningful classification of the flow biotope for river inventory, rehabilitation, design and appraisal purposes. Journal of Environmental Management 88: 638-650.

Harvey, G. L. and Clifford, N. J. (2008) Microscale hydrodynamics and coherent flow structures in rivers: implications for the characterisation of physical habitat, River Research and Applications, DOI: 10.1002/rra.1109.

Harvey, G. L. and Clifford, N. J. (in preparation) Field assessment of suspended sediment pathway experiments to characterise hydraulic habitat units in rivers.

Harvey, G. L. and Wallerstein, N. P. (in press) Assessing the Interactions between Flood Defence Maintenance Works and River Habitats: the Use of River Habitat Survey. *Aquatic Conservation: Marine and Freshwater Ecosystems.*

Heywood MJT Walling DE. 2007. The sedimentation of salmonid spawning gravels in the Hampshire Avon Catchment, UK: implications for the dissolved oxygen content of intergravel water and embryo survival. *Hydrological Processes* 71: 770-778.

Holmes, N. T. (1998) The River Restoration Project and its demonstration sites. In *Rehabilitation of Rivers: Principles and Implementation*, De Waal, L., Large, A. R. G. and Wade, P. M. (eds.). John Wiley and Sons: Chichester.

Hynes, H. B. N. (1970) The Ecology of Running Waters. Liverpool, Liverpool University Press.

Janauer, G. A. (2000) Ecohydrology: fusing concepts and scales. *Ecological Engineering* 16: 9-16.

Jansson, R., Backx, H., Boulton, A. J., Dixon, M., Dudgeon, D., Hughes, F. M. R., Nakamura, K., Stanley, E. H. and Tockner, K. (2005) Stating mechanisms and refining criteria for ecologically successful river restoration. *Journal of Applied Ecology* 42: 218-222.

Jeffers, J. N. R. (1998) Characterization of river habitats and prediction of habitat features using ordination techniques. *Aquatic Conservation: Marine and Freshwater Ecosystems* 8: 529-540.

Johnson et al (1995) Modelling the effect of groundwater abstraction on salmonid habitat availability in the river Allen, Dorset, England. *Regulated Rivers Research and Management* 10 (2-4): 229-238.

Jowett, I. G. (1993) A method for objectively identifying pool, run and riffle habitats from physical measurements. *New Zealand Journal of Marine and Freshwater Research* 27: 241-248.

Junk, W. J., Bayley, P. B. and Sparks, R. E. (1989) The flood pulse concept in river-floodplain systems. Canadian *Journal of Fisheries and Aquatic Sciences Special Publication* 106: 110-127.

Kaenel, B. R., Buehrer, H. and Uehlinger, U. (2000) Effects of aquatic plant management on stream metabolism and oxygen balance in streams. *Freshwater Biology* 45: 85-95.

Karr, J. (1999) Defining and measuring river health. *Freshwater Biology* 41: 221-234.

Kates, R. W., Clark, W. C., Corell, R., Hall, J. M., Jaeger, C. C., Lowe, I., McCarthy, J. J., Schellnhuber, H. J., Bolin, B., Dickson, N. M., Faucheux, S., Gallopin, G. C., Grubler, A., Huntley, B., Jager, J., Jodha, N. S., Kasperson, R. E., Mabogunje, A., Matson, P., Mooney, H., Moore, B. 3rd, O'Riordan, T., Svedlin, U. (2001) Environment and development: Sustainability science. *Science,* 292(5517), 641-2

Keller, E. A. (1976) Channelisation: environmental, geomorphic and engineering aspects. In *Geomorphology and Engineering*, Coates, D. R. (Ed.). Dowden Hutchinson & Ross: 115-40.

Kellerhals, R. and Miles, M. (1996) Fluvial geomorphology and fish habitat: implications for river restoration. Proceedings of the Second IAHR Symposium on Habitat Hydraulics, Ecohydraulics 2000, Quebec, INRS-Eau, FQSA, IAHR/AIRH.

Kershner, J. L. and Snider, W. M. (1992) Importance of a habitat-level classification system to design instream flow studies in: *River Conservation and Management* Boon, P. J., Calow, P. and Petts, G. E. (Eds.). John Wiley and Sons: Chichester, pp. 179-193.

Klein, L. R., Clayton, S. R., Alldredge, J. R. and Goodwin, P. (2007) Long-term monitoring and evaluation of the lower Red River meadow restoration project, Idaho, U.S.A. *Restoration Ecology* 15 (2): 223-239.

Kondolf, G. M. (1991) Hungry Water: Effects of Dams and Gravel Mining on River Channels. *Environmental Management* 21 (4): 533-551.

Kondolf, G. M., Boulton, A. J., O'Daniel, S., Poole, G. C., Rahel, F. J., Stanley, E. H., Wohl, E., Bang, A., Carlstrom, J., Cristoni, C., Huber, H., Koljonen, S., Louhi, P. and Nakamura, K. (2006) Process-based ecological river restoration: visualizing three-dimensional connectivity and dynamic vectors to recover lost linkages. *Ecology and Society* 11(2): 5

Kondolf, G. M. and Yang, C-N. (2008) Planning river restoration projects: social and cultural dimensions. In *River Restoration: Managing the Uncertainty in Restoring Physical Habitat*, Darby, S. and Sear, D. (eds). John Wiley and Sons Ltd: Chichester.

Kowaleswski, M., Avila Serrano, G. E., Flessa, K. W. and Goodfriend, G. A. (2000) Dead delta's former productivity: Two trillion shells at the mouth of the Colorado River. *Geology* 28 (12): 1059-1062.

Lancaster, J., Buffin-Belanger, T., Reid, I. and Rice, S. (2006) Flow- and substratum-mediated movement by a stream insect. *Freshwater Biology* 51: 1053-1069.

Lancaster, J. and Hildrew, A. G. (1993) Characterizing in-stream flow refugia. *Canadian Journal of Fisheries and Aquatic Sciences* 50: 1663-1675.

Lane, S. N., Mould, D. C., Carbonneau, P. E. and Hardy, R. J. (2006) Fuzzy modelling of habitat suitability using 2D and 3D hydrodynamic models: biological challenges. In *River Flow 2006: Proceedings of an International Conference on Fluvial Hydraulics, 6-8 September 2006,* Ferreira, R. M. L., Alves, E. C. T. L., Leal, J. G. A. B. and Cardoso, A. H. (Eds.). Taylor and Francis: London.

LeClerc, M. (2002) Ecohydraulics, last frontier for fluvial hydraulics: research challenges and multidisciplinary perspectives, *River Flow 2002*, Lisse, Swets & Zeitlinger.

Leopold, L B. (1969) Quantitative comparison of some aesthetic factors among rivers. US Geological Survey, Circular 620.

Leopold, Luna B., and Wolman, M.G. (1957) River Channel Patterns: Braided, Meandering and Straight, U.S. Geological Survey Professional Paper, 282-B.

Maddock, I. (1999) The importance of physical habitat assessment for evaluating river health. *Freshwater Biology* 41: 373-391.

Manchester, S. J. and Bullock, J. M. (2000) The impacts of non-native species on UK biodiversity and the effectiveness of control. *Journal of Applied Ecology* 37: 845-864.

Marsh, G. P. (1864) *Man and Nature; or, Physical Geography as Modified by Human Action.* Belknap Press of Harvard University: Cambridge, MA.

Millington, C. E. and Sear, D. A. (2007) Impacts of river restoration on small-wood dynamics in a low gradient headwater stream. *Earth Surface Processes and Landforms* 32: 1204-1218.

Moir, H. J., Gibbins, C. N., Soulsby, C. and Youngson, A. F. (2005) PHABSIM modelling of atlantic salmon spawning habitat in an upland stream: testing the influence of habitat suitability indices on model output. *River Research and Applications* 21: 1021-1034.

Mosely, M. P. (1982) River channel inventory, habitat and instream flow assessment. *Progress in Physical Geography* 9: 494-523.

Moss, B. (2000) Biodiversity in fresh waters - an issue of species preservation or system functioning? *Environmental Conservation* 27 (1): 1-4.

Mouton, A., Meixner, H., Goethals, P. L. M., De Pauw, N. and Mader, H. (2007) Concept and application of the usable volume for modelling the physical habitat of riverine organisms. *River Research and Applications* 23: 545–558

Muhar, S., Schmutz, S. and Jungwirth, M. (1995) River restoration concepts – goals and perspective. *Hydrobiologia* 303: 183–194.

Naiman, R. J., Lonzarich, D. G., Beechie, T. J. and Ralph, S. C. (1992) General principals of river classification and the assessment of conservation potential in rivers. In *River Conservation and Management*, Boon, P. J., Calow, P. and Petts, G. E. (eds.), John Wiley and Sons: Chichester.

Newson, M. D. (1992) *Land, Water and Development: River Basin Systems and their Sustainable Management.* Routledge: London.

Newson, M. D. (2002) Geomorphological concepts and tools for sustainable river ecosystem management. *Aquatic Conservation: Marine and Freshwater Ecosystems* 12: 365-379.

Newson, M. D. and Large, A. R. G. (2006) 'Natural' rivers, 'hydromorphological quality' and river restoration: a challenging new agenda for applied fluvial geomorphology. *Earth Surface Processes and Landforms*, 31:1606-1624.

Newson, M. D. and Newson, C. L. (2000) Geomorphology, ecology and river channel habitat; mesoscale approaches to basin-scale challenges. *Progress in Physical Geography* 24: 195-217.

Nielsen, M. B. (1996) River restoration: report of a major EU Life demonstration project. *Aquatic Conservation: Marine and Freshwater Ecosystems*, 6: 187-190.

Nienhuis, P. H. and Leuven, R. S. E. W. (2001) River restoration and flood protection: controversy or synergism? *Hydrobiologia* 444: 85-99.

O'Keefe, J. H. and Uys, M. (2000) The role of classification in the conservation of rivers. In *Global Perspectives in River Conservation*, Boon, P. J., Davies, B. R. and Petts, G. E. (eds.). John Wiley and Sons: Chichester.

Ormerod, S. J. (2004) A golden age of river restoration science? *Aquatic Conservation: Marine and Freshwater Ecosystems*, 14: 543-549.

Padmore, C. L. (1997) Physical biotopes in representative river channels: identification, hydraulic characterisation and application. Unpublished PhD Thesis, University of Newcastle upon Tyne.

Palmer, L. (1976) River Management criteria for Oregon and Washington. In *Geomorphology and Engineering*, Coates, D. R. (Ed.). Dowden Hutchinson & Ross: 329-46.

Palmer, C. G., Peckham, B. and Soltau, F. (2000) The role of legislation in river conservation. In Global Perspectives in River Conservation, Boon, P. J., Davies, B. R. and Petts, G. E. (eds.). John Wiley and Sons: Chichester.

Palmer, M. A., Berhardt, E. S., Allan, J. D., Lake, P. S., Alexander, G., Brooks, S., Carr, J., Clayton, S., Dahm, C. N., Follstad Shah, J., Galat, D. L., Loss, S. G., Goodwin, P., Hart, D. D., Hassett, B., Jenkinson, R., Kondolf, G. M., Lave, R., Meyer, J. L., O'Donnell, T. K., Pagano, L. and Sudduth, E. (2005) Standards for ecologically successful river restoration. *Journal of Applied Ecology* 42: 208-217.

Palmer, M. A., Allan, J. D., Meyer, J. and Bernhardt, E. S. (2007) River restoration in the twenty-first century: data and experiential knowledge to inform future efforts. *Restoration Ecology* 15 (3): 472-481.

Parasiewicz, P. (2007) Editorial: Overcoming the limits of scales. *River Research and Applications* 23: 891–892.

Parasiewicz, P. and Walker, J. D. (2007) Comparison of Mesohabsim with two microhabitat models (PHABSIM and HARPHA). *River Research and Applications* 23: 904–923

Park, C. C. (1981) Man, river systems and environmental impacts. *Progress in Physical Geography* 5: 1-31.

Petts, G. E. and Amoros, C. (1996) *Fluvial Hydrosystems*. Chapman and Hall: London.

Petts, G. E. and Greenwood, M. (1985) Channel changes and invertebrate faunas below Nant-Y-Moch dam, River Rheidol, Wales, UK. *Hydrobiologia* 122: 65-80

Petts, G. E., Maddock, I., Bickerton, M. and Ferguson, A. J. D. (1995) Linking hydrology and ecology: the scientific basis for river management. In *The Ecological Basis for River Management*, Harper, D. M. and Ferguson, A. J. D. (eds.), John Wiley and Sons Chichester: 1-16.

Piegay, H. Mutz, M. Gregory, K. J., Rinaldi, M., Bondarev, V., Wyzga, B., Chin, A., Zawiejska, J., Dahlstrom, N., Elosegi, A., Gregory, S. V. and Joshi, V. (2005) Public perception as a barrier to introducing wood in rivers for restoration purposes. *Environmental Management* 36 (5): 665-674.

Poff N. L., Allan, J. D., Bain, M. B., Karr, J. R., Prestegaard, K. L., Richter, B. D., Sparks, R. E. and Stromberg, J. C. (1997) The natural flow regime: A paradigm for river conservation and restoration. *Bioscience* 47 (11): 769-784.

Poff, N. L., Allan, J. D., Palmer, M. A., Hart, D. D., Richter, B. D., Arthington, A. H., Rogers, K. H., Meyer, J. L. and Stanford, J. A. (2003) River flows and water wars: emerging science for environmental decision making. *Frontiers in Ecology and the Environment* 1 (6): 298–306.

Poudevigne, I., Alard, D., Leuven, R. S. E. W. and Nienhuis, P. H. (2002) A systems approach to river restoration: a case study in the lower Seine valley, France. *River Research and Applications* 18: 239-247.

Pretty, J. L., Harrison, S. S. C, Shepherd, D. J., Smith, C., Hildrew, A. G. and Hey, R. D. (2003) River rehabilitation and fish populations: assessing the benefit of instream structures. *Journal of Applied Ecology*, 40: 251-265.

Raven, P. J., Fox, P., Everard, M., Holmes, N. T. H. and Dawson, F. H. (1997) River habitat survey: a new system for classifying rivers according to their habitat quality. In *Freshwater Quality: Defining the Indefinable?*, Boon, P. J. and Howell, D. L. (eds.). The Stationery Office: Edinburgh. P. 215-234.

Rhoads, B. L. and Herricks, E. E. (1996) Naturalization of headwater streams in Illinois: challenges and possibilities. In *River Channel Restoration*, Brookes, A. and Shields, F. D. (Eds.). John Wiley & Sons: Chichester.

Rhoads, B. L., Wilson, D., Urban, M. and Herricks, E. E. (1999) Interaction between scientists and nonscientists in community-based watershed management: emergence of the concept of stream naturalization. *Environmental Management* 24 (3): 297-308.

Rhoads, B. L., Garcia, M. H., Rodriguez, J., Bombardelli, F., Abad, J. and Daniels, M. (2008) Methods for evaluating the geomorphological performance of naturalized rivers: examples from the Chicago metropolitan area. In *River Restoration: Managing the Uncertainty in Restoring Physical Habitat*, Darby, S. and Sear, D. (Eds). John Wiley and Sons Ltd: Chichester.

Richards, K. S. (1996) Samples and cases: generalisation and explanation on Geomorphology. The Scientific Nature of Geomorphology: Proceedings of the 27th Binghamton Symposium in Geomorphology, Binghamton, USA, John Wiley & Sons Ltd.

Roni, P., Beechie, T. J., Bilby, R. E., Leonetti, F. E., Pollock, M. M. and Pess, G. R. (1992) A Review of Stream Restoration Techniques and a Hierarchical Strategy for Prioritizing Restoration in Pacific Northwest Watersheds. *North American Journal of Fisheries Management* 22:1–20.

Rosgen, D. L. (1994) A Classification of Natural Rivers, *Catena* 22: 169-199.

Rosgen, D.L., 1996. Applied River Morphology. Wildland Hydrology Books, Pagosa Springs, Colorado, and Ft. Collins, CO.

Sayer, A. (1992) Method in Social Science: A Realist Approach. London, Routledge.

Schumm, S.A., (1985) Patterns of alluvial rivers. *Annual Review of Earth and Planetary Sciences* 13: 5–27.

Sear, D. A., Briggs, A. and Brookes, A. (1998) A preliminary analysis of the morphological adjustment within and downstream of a lowland river subject to river restoration. *Aquatic Conservation: Marine and Freshwater Ecosystems*, 8: 167-183.

Shaler, N. S. (1905) *Man and the Earth*. The Chautauqua Press: New York.

Shields, F. D., Cooper, C. M. and Knight, S. S. (1993) Initial habitat response to incised channel rehabilitation. *Aquatic Conservation: Marine and Freshwater Ecosystems* 3: 93-103.

Showers, K. B. (2000) Popular participation in river conservation. In *Global Perspectives in River Conservation*, Boon, P. J., Davies, B. R. and Petts, G. E. (eds.). John Wiley and Sons: Chichester.

Skinner, K. S. and Bruce-Burgess, L. (2005) Strategic and project level river restoration protocols: key components for meeting the requirements of the Water Framework Directive (WFD). Journal of the Chartered Institution of Water and Environmental Management, 19(2): 135-142.

Smith, C., Youdan, T. and Redmond, C. (1995) Practical aspects of restoration of channel diversity in physically degraded streams. In *The Ecological Basis for River Management*, Harper, D. M. and Ferguson, A. J. D. (eds.). John Wiley and Sons: Chichester. P. 269 - 273.

Soar, P. J. and Thorne, C. R. (2001) Channel restoration design for meandering rivers. US Army Corps of Engineers Engineering and Research Development Centre: Coastal and Hydraulics Laboratory Report ERDC/CHL CR-01-1.

Soulsby C, Youngson AF, Moir HJ, Malcolm IA. 2001. Fine sediment influence on salmonid spawning habitat in a lowland agricultural stream: a preliminary assessment. *The Science of the Total Environment* 265: 295-307.

Southwood, T. R. E. (1977) Habitat, the templet for ecological strategies? *Journal of Animal Ecology* 46(2): 336-365.

Spence, R. and Hickley, P. (2000) The use of PHABSIM in the management of water resources and fisheries in England and Wales. *Ecological Engineering* 16: 153-158.

Stanley, E. H. and Boulton, A. J. (2000) River size as a factor in river conservation. In *Global Perspectives in River Conservation*, Boon, P. J., Davies, B. R. and Petts, G. E. (eds.). John Wiley and Sons: Chichester.

Stoll-Kleeman, S. and O'Riordan, T. (2002) Enhancing biodiversity for humanity. In *Biodiversity, Sustainability and Human Communities: Protecting Beyond the Protected*, Stoll-Kleeman, S. and O'Riordan, T. (Eds.). Cambridge University Press.

The European Parliament and the Council of the European Union (2000) Directive 2000/60/EC of the European Parliament and of the Council establishing a framework for the Community action in the field of water policy.

The European Parliament and the Council of the European Union. (2007) Directive 2007/60/EC of the European Parliament and of the Council on the assessment and management of flood risks.

Thomas, W. L. (ed.) (1956) *Man's Role in Changing the Face of the Earth.* University of Chicago Press: Chicago.

Tickner, D., Armitage, P. D., Bickerton, M. A. and Hall, K. A. (2000) Assessing stream quality using information on mesohabitat distribution and character. *Aquatic Conservation: Marine and Freshwater Ecosystems* 10: 179-196.

Townsend, C. R. (1989) The patch dynamics concept of stream community ecology. *Journal of the North American Benthological Society* 8(1): 36-50.

Townsend, C. R. (1996) Concepts in river ecology: pattern and process in the catchment hierarchy. *Archiv fur Hydrobiologie Supplement 113, Large Rivers* 10: 3-21.

Tsujimoto, T. (1996) Fish habitat and micro structure of flow in gravel bed stream. Proceedings of the Second IAHR Symposium on Habitat Hydraulics, Ecohydraulics 2000, Quebec, INRS-Eau, FQSA, IAHR/AIRH.

Tunstall, S. M., Penning-Rowsell, E. C., Tapsell, S. M. and Eden, S. E. (2000) River restoration: public attitudes and expectations. *Water and Environment Journal* 14: 363-370.

Udvardy, M. F. D. (1959) Notes on the ecological concepts of habitat, biotope and niche. *Ecology* 40(4): 725-728.

Van Rijen, JPM (1998) Practical approaches for nature development: let nature do its own thing again. In *Rehabilitation of Rivers: Principles and Implementation*, De Waal, L., Large, A. R. G. and Wade, P. M. (eds.). John Wiley and Sons: Chichester.

Vannote, R. L., Minshall, G. W., Cummins, K. W., Sedell, J. R. and Cushing, C. E. (1980) The River Continuum Concept. *Canadian Journal of Fisheries and Aquatic Sciences* 37: 130-137.

Vaughan, I., Diamond, M., Gurnell, A. M., Jenkins, A., Milner, N. J., Naylor, L., A., Sear, D. A., Woodward, G., and Ormerod, S. J. (2007) Integrating ecology with hydromorphology: a priority for river science and management. *Aquatic Conservation: Marine and Freshwater Ecosystems*. DOI: 10.1002/aqc.895.

Wadeson, R. A. (1994) A Geomorphological approach to the identification and classification of instream flow environments. *South African Journal of Aquatic Sciences* 20(1/2): 38-61.

Walsh, C. J., Roy, A. H., Feminella, J. W., Cottingham, P. D., Groffman, P. M. and Morgan II, R. P. (2005) The urban stream syndrome: current knowledge and the search for a cure. *Journal of the North American Benthological Society* 24 (3): 706-723.

Ward, J. V. and J. A. Stanford (1983). The serial discontinuity concept of lotic ecosystems. In *Dynamics of Lotic Ecosystems,* Fontaine, T. D. and Bartell, S. M. (eds.). Ann Arbor Science: 29-41.

Wharton, G. and Gilver, D. J. (2006) River restoration in the UK: meeting the dual needs of the European Union Water Framework Directive and flood defence? *International Journal of River Basin Management* 4 (4): 1-12.

Whalen, P. J., Toth, L. A., Koebel, J. W., Strayer, P. K. (2002) Kissimmee River restoration: a case study. *Water Science and Technology* 45 (11): 55-62.

Wheaton, J. M. (2005) Review of River Restoration Motivations and Objectives. Unpublished Review, Southampton, UK. 12pp.

Wohl, E., Angermeier, P. L., Bledsoe, B., Kondolf, G. M., MacDonnell, L., Merritt, D. M., Palmer, M. A., Poff, N. L. and Tarboton, D. (2005) River restoration. *Water Resources Research* 41: W10301.

Wishart, M. J., Davies, B. R., Boon, P. J. and Pringle, C. M. (2000) Global disparities in river conservation: 'First World' values and 'Third World' realities. In *Global Perspectives in River Conservation*, Boon, P. J., Davies, B. R. and Petts, G. E. (eds.). John Wiley and Sons: Chichester.

Woolsey, S., Capelli, F., Gonser, T., Hoehn, E., Hostmann, M., Junker, B., Paetzold, A., Weber, C. and Peter, A. (2007) A strategy to assess river restoration success. *Freshwater Biology* 52: 752-769.

In: Ecological Restoration
Editors: George H. Pardue and Thomas K. Olvera

ISBN 978-1-60741-013-3
© 2009 Nova Science Publishers, Inc.

Chapter 3

EFFECTS OF ECOLOGICAL RESTORATION ON BIOGEOCHEMICAL PROCESSES IN COMPLEX RIVERINE LANDSCAPES

Thomas Hein[1,2]*, Elisabeth Bondar-Kunze[1,2], Friedrich Schiemer[3], Nina Welti[1,2] and Gilles Pinay[4]

[1] University for Natural Resources and Applied Life Sciences, Institute of Hydrobiology and Aquatic Ecosystem Management, Vienna, Austria
[2] WasserKluster Lunz GmBH, Interuniversity Center for Aquatic Ecosystem Research, Lunz/See, Austria
[3] University of Vienna, Dept. of Freshwater Ecology, Vienna, Austria
[4] School of Geography, Earth and Environmental Sciences, University of Birmingham, Edgbaston, Birmingham, United Kingdom

ABSTRACT

The deterioration of the functioning of river ecosystems has prompted rehabilitation and restoration measures during the last 20 years, based on the assessment of the ecological integrity. Most of these efforts have increased the spatial heterogeneity of these ecosystems. Yet, a more integrated approach including restoration of landscape dynamics and key ecosystem processes such as carbon and nutrient cycling is necessary. Large-scale rehabilitation and restoration projects, therefore, also consider altered nutrient dynamics or aim at reducing nutrient transport in river corridors by increasing nutrient retention. The present chapter analyzes the effects of river side-arm restoration on ecosystem functions within the side-arm, and highlight potential effects on the main channel and downstream areas in a large river. We demonstrated that principles of hydromorphological dynamics control nutrient cycling, phytoplankton production and interactions within the planktonic food web. These findings confirm the environmental control on these biological processes and their potential use as proxies to assess the consequences of hydrological changes restoration measures on river ecosystem functioning.

Specifically, we addressed the following questions related to ecological restoration of side arms: (i) what are the effects of varying hydrological connectivity on nutrient concentrations and the retention capacity?; (ii) can easily derived hydrological parameters provide meaningful information on algal development and planktonic food web interactions?; and (iii) what implications do these predictions have for future rehabilitation work and integrating these effects on larger scales (sub-catchment)?

INTRODUCTION

Riverine ecosystems control the transport of nutrients and organic matter from terrestrial sources (Bennett et al. 2001; Seitzinger et al. 2002b; Townsend-Small et al. 2005), produce organic material within aquatic environments, degrade organic matter while transporting it downstream (Hedges et al. 2000) and carry the fingerprint of human activities (Rosenberg et al. 2000). The efficiency of nutrient transformation versus transport is controlled by the hydrological dynamics as well as the spatial complexity of the riverine ecosystem (Hein et al. 2005). Therefore, floodplains, riparian zones and instream structures, by increasing the complexity of the river system at the landscape level, are key components of controlling these functions within riverine landscapes (Hynes 1975; Naiman & Décamps 1997; Ren et al. 2000) and act as biogeochemical hot spots (Balser et al. 2006; Forshay & Stanley 2005; McClain et al. 2003).

Large rivers with relatively complex morphological structures and hydrological exchange patterns have the potential for an intense turnover of organic matter and inorganic solutes due to high algal and microbial activity (Fischer et al. 2005). Analysis of biogeochemical budgets indicates that river networks can remove 37–76% of the total N-input, mainly via denitrification, with a high contribution by high-order river sections (Seitzinger et al. 2002a; Seitzinger et al. 2002c). Large rivers are therefore important for the biogeochemical budgets of catchments (Behrendt & Opitz 2000; Seitzinger et al. 2002a), even though the water depth-related retention in river channels decreases along a river continuum (Alexander et al. 2000; Allan 1995). Large rivers also process high amounts of organic carbon (Fischer et al. 2002; Hedges et al. 2000), playing a crucial role in the carbon cycling of estuarine and coastal regions (Raymond & Bauer 2001).

More intense matter cycling in these landscape elements is related to the mode of water exchange and their water retention capacity (Carling 1992) which impact river water quality at the larger scale (Brinson 1993; Hefting et al. 2006; Peterjohn & Correll 1984; Pinay & Decamps 1988; Triska et al. 1989). These retention structures range spatially from in-channel zones to terrestrial components of floodplains and temporally from frequently to rarely connected water bodies. Retention areas, defined as slackwater areas (Thorp & Casper 2003), frequently connected to the riverine flow, substantially support the local aquatic production and the associated nutrient transformation at water levels below bankfull and low order flood events (Schiemer et al. 2001; Weilhoefer et al. 2008). The distance to the main channel and the regularity of connectivity explain the duration and frequency of nutrient pulses to a certain slackwater area during the year (Tockner et al. 2000). The lateral dimension of slackwaters governs the overall retention capacity of individual water bodies within the floodplain (Heiler et al. 1995), but the exchange between surface and subsurface also contributes significantly to the nutrient retention capacity (Richardson 2004). Indeed, they also represent functional retention areas (Carling, 1992) which control and maintain river water quality (Peterjohn & Correll 1984; Brinson et al. 1983; Pinay & Décamps 1988; Triska et al. 1989; Hefting et al.

2006). In an unregulated anabranched section, a sequence of retention areas with decreasing duration and frequency of surface connectivity can typically be found along the lateral extension of the river corridor (Fig. 1).

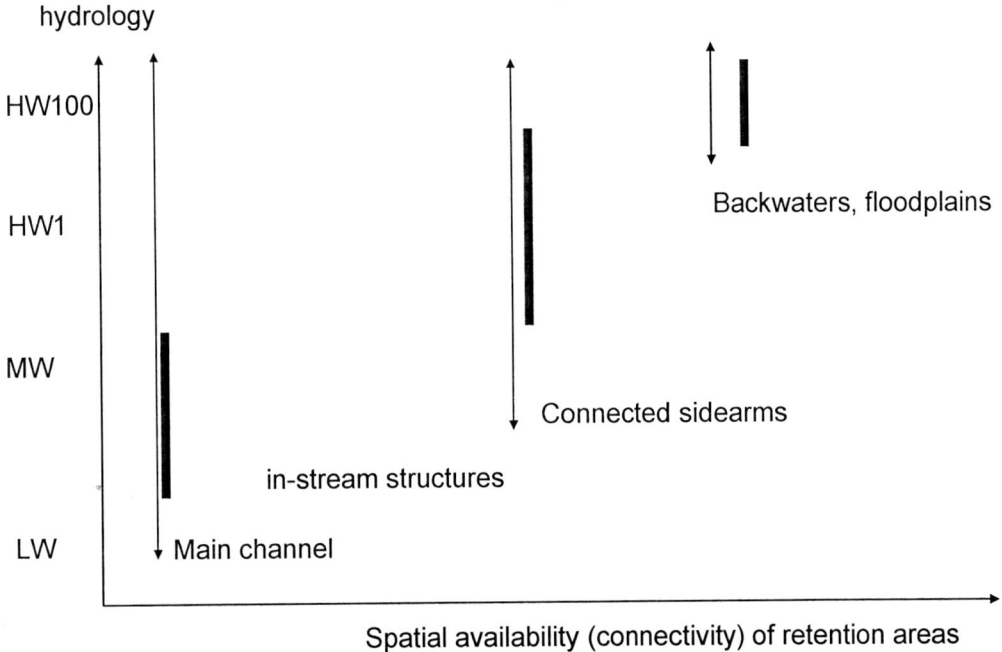

Figure 1. Schematic representation of potential retention areas in a braided river reach. Arrows indicate surface water connectivity with the main channel, dark bars the theoretical time windows during which they act as a retention area (increased water retention time compared to the mean travel time in the main channel).

The actual performance of these retention areas depends on the exchange conditions (the connectivity) and the morphological setting in the respective river corridor. At the landscape scale, three interrelated principles of hydromorphological dynamics can be formulated regarding the cycling and transfer of carbon and nutrients in large rivers ecosystems: i) the mode of carbon and nutrient delivery affects ecosystem functioning; ii) increasing residence time and contact area impact nutrient transformation by biota; iii) floods and droughts are natural events that strongly influence pathways of carbon and nutrient cycling (Pinay et al. 2007). This interaction between hydromorphology and matter cycling provides the basis for understanding biotic development in terms of biomass (Basu & Pick 1997), species diversity (Tockner et al. 2006) and food web relationships (Woodward et al. 2005).

The first principle is related to the mode of carbon and nutrient inputs controlled by the flow regime along river ecosystems. River systems and their retention zones can be viewed as open ecosystems dynamically linked longitudinally, laterally and vertically by hydrologic and geomorphologic processes (Ward 1989). In floodplains of most large rivers, the input of nutrients, sediment and organic matter occur mainly via upstream surface flow. Significant amounts of these materials are deposited during floods (Brinson et al. 1983; Brunet et al. 1994; Grubaugh & Anderson 1989; Lowrance et al. 1986; Van der Lee et al. 2004). The transfer and storage of materials in floodplains are largely under the control of the connectivity pattern within the river landscape as well as of the magnitude, frequency and duration of floods.

Hydrological connectivity includes not only the lateral extent, but also the vertical component. Especially in braided river sections, this calls for additionally considering the spatial and temporal complexity of subsurface flows (Tockner et al. 2006). This exchange pattern influences benthic processes related to the surface such as algal primary production and nutrient uptake in the root zone in deeper layers and organic matter degradation along a gradient of oxygen availability, affecting the overall stability of the riverine landscapes (Fisher et al. 1998). Hydrological connectivity creates a mosaic of geomorphologic surfaces that influence the spatial distribution and successional development of different communities (Roberts & Ludwig 1991; Salo et al. 1986) In dynamically connected retention areas, the pattern of surface and sub-surface flow provides the basis for intensive nutrient re-cycling (Pinay et al. 2007).

The second basic principle of hydromorphological dynamics is that increasing residence time as well as increasing contact zones in retention areas (i.e., water-sediment contact) is positively correlated to the efficiency of nutrient transformation in river ecosystems. These positive relationships occur both in the main channel itself and in the riparian and floodplain zones (Hill 1979; Jones & Holmes 1996; Ponnamperuma 1972; Valett et al. 1996). Increased residence times are crucial to understanding the decomposition of organic matter (Pusch et al. 1998) and the development of plankton biota in large river systems (Basu & Pick 1997; Reckendorfer et al. 1999; Weitere & Arndt 2002a). The morphology – the spatial heterogeneity (complexity), the duration of contact between water and surfaces – influences the biological use of nutrients and thereby the total amount of nutrients processed (Gergel et al. 2005). For example, empirical evidence shows that the rate of instream nitrogen cycling depends on the surface contact of water with sediments and their organic content as well as on the frequency and duration of contact in retention areas (McClain et al. 2003). The locally produced organic matter has the potential to subsidize main channel communities in the respective section and in downstream sections (Junk & Wantzen 2004; Tockner et al. 2006).

The third principle is related to the role of rapidly changing water levels, especially flow and flood pulses (Tockner et al. 2000), in shaping the characteristics of carbon and nutrient cycling in time and, hence, the resilience of the system responding to pulsed input and disturbance events. Thus, water regime or connectivity changes directly affect – by altering either the frequency, duration, period of occurrence or intensity of water levels – the nutrient export at the catchment scale. (Alvarez-Cobelas et al. 2008; Behrendt et al. 2005) and at the reach scale in varying retention areas (e.g. Baker & Vervier 2004; Hein et al. 2004b). The extent, composition and configuration of aquatic and terrestrial habitats vary in response to the pulsing of discharge. Flooding duration is controlled by the hydrograph and local topography; low areas are flooded more often and longer than higher ones, producing variations in biogeochemical patterns at the meter scale (Hogan et al. 2004; Pinay et al. 1989; Pinay & Naiman 1991; Wright et al. 2001).

In large rivers the pulsing surface connectivity affects the retention areas, especially slackwater areas, and provides the basis for enhanced aquatic primary production (Ahearn et al. 2006; Schemel et al. 2004). Hence, the aquatic organic matter resources subsidizes main channel communities in arid zone rivers (Bunn et al. 2003) and leads to pulsed resource availability in temperate ones (Delong et al. 2001; Hein et al. 2003).

These three principles of hydromorphological dynamics build up the basis for the presented research in this chapter and are linked over different temporal and spatial scales (Fig. 2). All three factors can be strongly affected by natural disturbances or anthropogenic impacts, through a change in either the water regime or the geomorphologic setting of the river valley (Gergel et al. 2005). Any change in natural water regimes will affect the biogeochemistry of riparian and instream zones as well as their ability to cycle and mitigate nutrient fluxes originating from upstream and/or upslope.

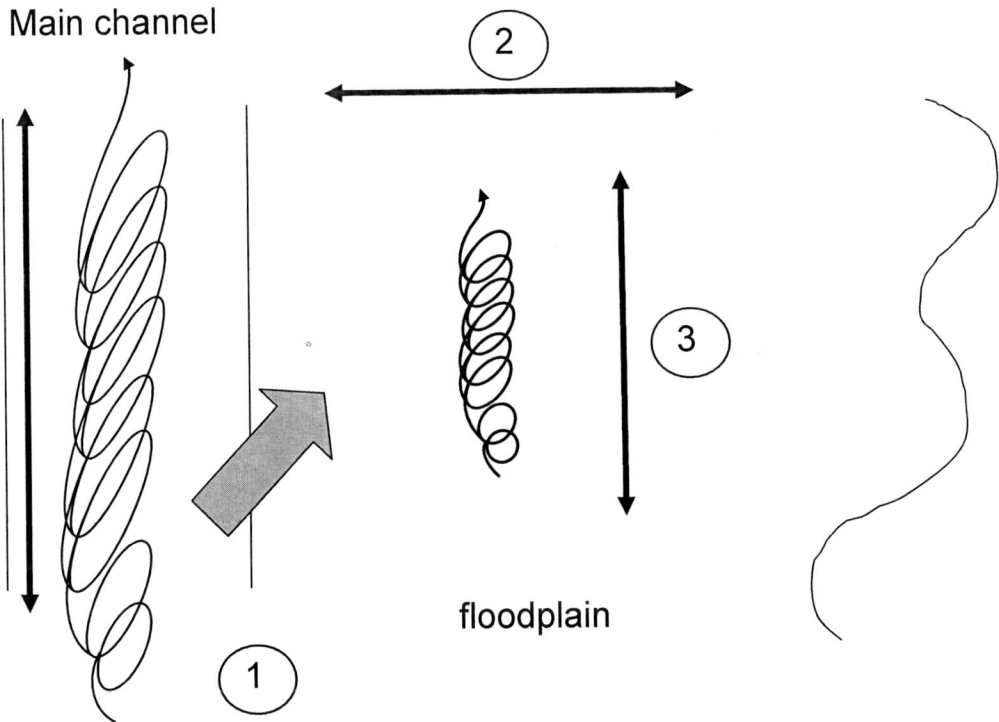

Figure 2. Schematic representation of the 3 hydrogeomorphic principles for biogeochemical functioning of floodplains. Spirals represent the transformation capacity in the main channel and adjacent floodplain. Numbers indicate the principles mentioned in the introduction, arrows present the variability of these principles in space and time. For principle 1 the extent of matter supply, for principle 2 the increase of surface, the spatial variability within the floodplain and for principle 3 the response on hydrological pulses and varying transformation capacity is shown.

Human Impacts on Complex River Ecosystems

Human populations have settled in river environments since prehistoric times, despite the obvious hazards. Adaptation of rivers and floodplains for habitation, agriculture and navigation also has a long history and even dates back to medieval times. In the course of the 20th century, many large rivers have undergone large-scale civil engineering operations. Thus, riverine ecosystems have been severely altered throughout the world, where for example, braided rivers are among the most endangered ecosystems globally (Sadler et al. 2004). Large rivers and their – often extensive – floodplains are among the most productive and diverse ecosystems on our planet (Tockner & Stanford 2002).These human activities at various scales have markedly altered the environmental conditions impacting the principles of hydromorphology described above and therefore the biogeochemical cycling in riverine landscapes.

At the catchment to continental/global scale, the temperature increase for the 21st century is predicted to range between 1.4 °C and 4.7 °C (Richards 1993). As the Earth warms, the hydrological cycle will be affected, likely intensifying hydrological cycling (Miller & Russell 1992) with severe impacts on the nutrient cycling as shown for the upper Danube River (Zweimüller et al. 2008). Precipitation, evapotranspiration and runoff are all expected to increase together with extreme events, i.e. floods and droughts (Loaiciga et al. 1996). In Europe, future scenarios of altered hydrological regime forecast an overall increase in the inter-annual variability of runoff (Arnell 1999) especially in the eastern part of the continent. The intensity, timing, frequency and duration of extreme hydrologic events are essential for maintaining the mosaic of habitats (Benke et al. 2000) as well as for the hot spots of processing and transformations in varying retention areas of river ecosystems (Junk et al. 1989; Naiman & Décamps 1997).

Numerous anthropogenic activities (e.g. dams and levees) have led to river ecosystem fragmentation and habitat destruction, disrupting the structures and functions of lotic ecosystems (Friedl & Wüest 2002; Gergel et al. 2005):

- Land use change has altered the magnitude, quality and timing of nutrient inputs into riverine systems.
- Flood control, electric power generation, navigation, or recreation impoundments change flow patterns, alter nutrient and sediment loads, and modify water temperatures.
- Regulation, canalization and flood protection can also significantly reduce the retention capacities of the riverine landscape and limit the exchange of matter to short periods of high flow (Tockner et al. 1999).

Within the Danube River basin, aerial estimates demonstrate that about 80% of the former floodplain areas are lost today (WWF 1999). The remaining areas show a distinct decline of ecosystem function due to restricted hydrologic exchange, as evidenced by the stretch of the Danube River downstream of Vienna. Reduced retention areas combined with increased nutrient inputs have significantly decreased the nutrient and organic matter retention capacity along the river ecosystem; the outputs have even altered coastal ecosystems (Schmidt 2001).

The deterioration of the functioning of river ecosystems has prompted rehabilitation and restoration measures in the last 20 years, based on the assessment of the ecological integrity (e.g. Jungwirth et al. 2002). Most measures have aimed at increasing the spatial heterogeneity of these ecosystems (Henry et al. 2002). Yet, a more integrated approach including restoration of landscape dynamics and key ecosystem processes such as carbon and nutrient retention is necessary (Canavan et al. 2007; Hohensinner et al. 2004; Pedroli et al. 2002). Large-scale rehabilitation and restoration projects must therefore also consider altered nutrient dynamics (Buijse et al. 2002) or aim at reducing nutrient transport in river corridors by increasing nutrient retention (Dahm et al. 1995; Mitsch & Day 2004).

The presented results in this chapter contributed novel results to the previously described principles and demonstrate the effects of changes in these principles related to ecological restoration on nutrient dynamics, phytoplankton primary productivity and contribution to organic matter sources and species diversity in plankton communities. Research results were presented of a case study of a river floodplain stretch along the Austrian Danube River downstream of Vienna.

THE AUSTRIAN DANUBE STRETCH DOWNSTREAM OF VIENNA

The Danube River is one of the main drainage systems in Europe (817,000 km^2). In Austria, the Danube is a 9th order river with a drainage basin of 104 000 km^2. The flow is characterized by an alpine regime with highly variable and stochastic patterns. The mean discharge is 1900 m^3 s^{-1} (Q$_{95}$: 950 m^3 s^{-1}, Q$_1$: 5040 m^3 s^{-1}). Like all large rivers in the industrialized world (Petts et al. 1989; Ward 1998) the ecology of the Danube River has been considerably affected by land-use changed, by pollution and most importantly by hydro-engineering (Schiemer et al. 1999).

Originally, the side-arms along the upper Danube stretch downstream of Vienna presented lotic conditions almost throughout the year, exhibiting a gradient from low to high flow velocities. This natural river system, a typical braided stretch, was relatively shallow and characterized by unstable banks. Large-scale floods occurred in irregular intervals and led to permanent channel migrations as well as to the formation of new channels, gravel banks and islands (Reckendorfer et al. 2005). High sediment dynamics created large alluvial fans in this unconstrained section, with floodplains several km wide (e.g. Schiemer et al. 2001). After the major regulation scheme of 1875, long term development lead to reduced hydrologic connectivity and dramatic loss of riverine habitats. The high nature value and potential for restoring key ecological processes led to the declaration of a national park in 1996. The present biogeochemical dynamics in the Danube are affected by the reduced quality and quantity of inshore zones and the limited lateral integration of former side-arms (Hein et al. 2005). Efforts are now being taken to improve the geomorphologic, hydrologic and ecologic conditions in reaction to the present situation (Reckendorfer et al. 2005). One aim is the re-establishment of the former continuous upstream connection as shown in the scheme (Fig. 3).

The Restoration Scheme for This Floodplain River Stretch

Fluvial dynamics are the most important driving forces for the development of long-term, self-sustaining alluvial river landscapes that exhibit a high degree of biodiversity (Ward et al. 2002). In the Alluvial Zone National Park, the side arms originally showed lotic conditions almost throughout the year. Areas with low flow velocities existed only in bays, abandoned arms and in side arms far away from the main stem. Today these former side-arms vary greatly in connectivity and differ in their rehabilitation potential. The main goal of the restoration programmes in the Alluvial Zone National Park is to increase the upstream surface connection with the river and the connectivity within the side-arm to enhance the duration of flowing conditions (Fig. 3, 4). The enhancement of connectivity with the Danube main channel was first established by lowering the riverside embankments and by additional artificial openings in different inflow areas (Schiemer et al. 1999).

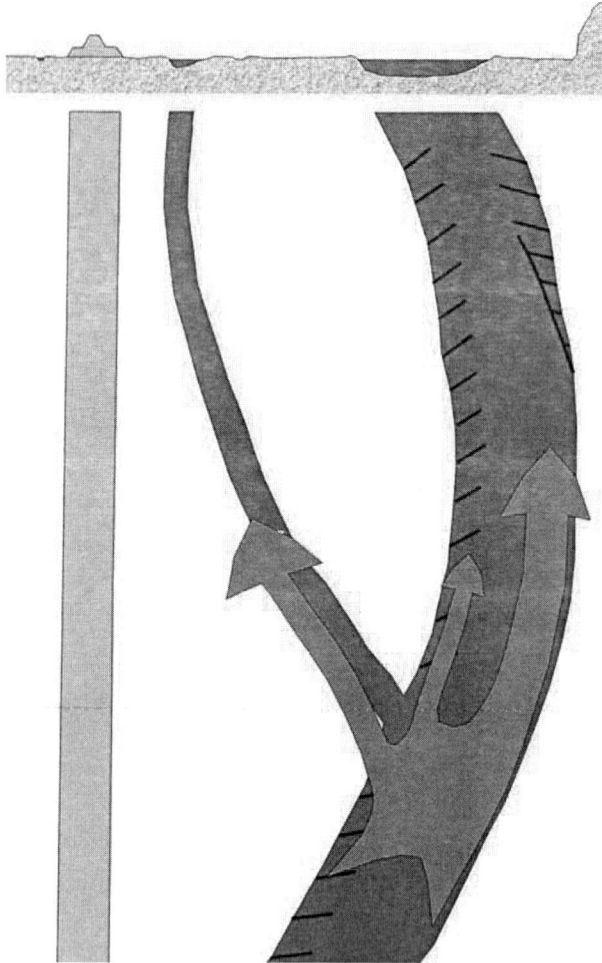

Figure 3. Schematic representation of side arm reconnection in anabranched river stretches (after Reckendorfer et al. 2005). Courtesy of C. Baumgartner.

The Regelsbrunn area, the first demonstration site of the Danube Restoration Programme (Hein et al. 1999; Tockner et al. 1998), was characterized by high potential of hydrologic exchange prior to rehabilitation (Heiler et al. 1995). The restoration programme carried out in the late 1990s, included the re-opening of the side-arm at 6 inlets over the entire stretch of 10 km and a lowering of the check-dams and additional culverts within the system, to approach pristine conditions (Fig. 4). A few years later a second site, the side-arm system Orth, was restored. The restoration programme for the Orth section of the Danube represents the second restoration step by building larger and deeper inflow areas that resulted in a surface connection in the range of 46 to 160 d a^{-1} (Hein et al. 2004b). The connectivity was increased between mean water and bankfull level at both sites. The proportion of discharge passing through Regelsbrunn after rehabilitation increased non-linearly with riverine flow conditions, ranging from less than 0.5 % at low water (< 6 m^3 s^{-1}) up to 12 % (about 650 m^3 s^{-1}) at high water. The restored sites were compared with the Lobau semi-isolated floodplain site, which is characterized by low connectivity levels and a downstream connection to the main river channel (Hein et al. 2004b).

Figure 4. Map of the restoration measures in the floodplain area of Regelsbrunn. © Nationalpark Donau-Auen and OeBF group. Insert: Drawings of technical measures for the inflow areas in detail, courtesy of K. Tockner.

A detailed hydrological model was developed to help quantifying the changes in ecological processes (Reckendorfer & Steel 2004). The software program 'Regels' (Reckendorfer & Steel 2004) was used to calculate water inflow and various hydrological and morphological parameters, such as basin volume, water surface area, mean flow velocity and discharge within the side-arm. The program also computed an implicit groundwater infiltration rate, depending on water levels in the side-arm, the main channel and the distance to the main channel. The model output metric "water age", which described the residence time adapted to the multi-input system, explained to a high degree nutrient uptake and pelagic processes in the water column of the reopened side-arm (Aspetsberger et al. 2002; Baranyi et al. 2002; Hein et al. 2003). Low age in the side-arm implies conditions similar to those in the main channel. The longer the water remained in the side-arm, the more that biological processes in the open-water and benthic systems could affect water quality. Hence, water age is an inverse measure of the connectivity to the main channel, with low age indicating high connectivity. A part of the ecological evaluation studies focused on changes in biogeochemical processes (Hein et al. 2003).

EFFECTS ON BIOGEOCHEMICAL PROCESSES
IN RIVERINE LANDSCAPES

Changes in surface hydrological connectivity impact biogeochemical cycling in side-arm systems. The principles of hydromorphological dynamics (see introduction) are directly related to the extent of nutrient transformation taking place in active side-arm systems, as the supply of nutrients and availability of water bodies and the retention time control the nutrient transformation. Hein et al. (2004b) describes in detail how connectivity influences the geochemical, nutrient and particle concentrations in restored side-arms. According to the type of water body (see also Hein et al. 2004a), mean nutrient and particle concentrations increased with connectivity, whereas geochemical values decreased.

For nutrients and particles the mean values approached riverine concentrations with increased surface water connectivity as also shown for other floodplain areas (Weilhoefer et al. 2008). For organic particles and particulate phosphorus concentration a decrease of concentration was predicted at medium connectivity values. Higher concentrations of particulate organic matter were measured during low water conditions when internal processes were important and during high water period when transport of allochthonous particles prevailed.

The changes over the connectivity gradient depended largely on the type of response function for each parameter. Small modifications in connectivity patterns led to different evolutions for dissolved, particulate and more process-related parameters (e.g. chlorophyll a). While changes in dissolved nutrients were already evident with small changes at low connectivity levels, mean particle concentrations linearly increased over the whole range of connectivity levels. Furthermore, connectivity influenced also the variability of chemical values in side-arms. The variability of geochemical values decreased with increasing connectivity, while that of phosphorus and particles increased.

Hein et al. (2005) discussed the role of retention areas for the whole riverine landscape. Two examples of the regulated, but still free-flowing section of the Danube River in Austria showed the effects on nutrient dynamics for a connected side-arm (using a budget approach). In contrast to nitrate, the side-arm very efficiently retained phosphorus during floods when the terrestrial component of the floodplain was inundated (Fig. 5 adapted from Coops et al. 2006).

During discharge conditions below bankfull the retention and transformation capacity of dissolved phosphorus was more pronounced in the side-arm system. Comparing the retention, removal and production rates of the studied side-arm showed the importance of the side-arm for nitrate removal during low flows, for algal production during medium flows and for phosphorus retention during high flows (Fig. 5).

Based on this budgetary approach, we demonstrated that at specific discharge conditions – varying time windows – different functions in the restored side-arm system have been activated in relation to the supply of nutrients and the residence time in the side-arm system (principle 1 and 2 as proposed in introduction).

Comparing different hydrological years based on this budget approach clearly illustrated the high inter-annual variability of the retention capacity and the high variation between different compounds, in this case total phosphorus and nitrate (Table 1).

Table 1. Discharge characteristics of the main channel of the Danube River (DR) and days of flooding (days with discharge higher than HQ1. Annual sum of phosphorus retention and nitrate removal for the Regelsbrunn area. Basis for the annual phosphorus retention (expressed in phosphorus) and nitrate removal estimation (expressed in nitrogen) was the relationship presented in Fig. 4, for details see text. Q: discharge in $m^3 s^{-1}$; STD: standard deviation.

Year	Mean Q DR $m^3 s^{-1}$	STD Q DR $m^3 s^{-1}$	Days of flooding	Nitrate removal $mt\ y^{-1}$	Phosphorus retention $t\ y^{-1}$
1997	2010	814	8	423.21	42.89
1998	1967	590	1	606.85	11.20
1999	2292	856	8	497.38	33.94
2000	2259	678	0	456.31	17.79
2001	2148	666	1	394.23	16.06
2002	2481	1176	12	328.45	174.62
2003	1632	473	0	653.23	3.43

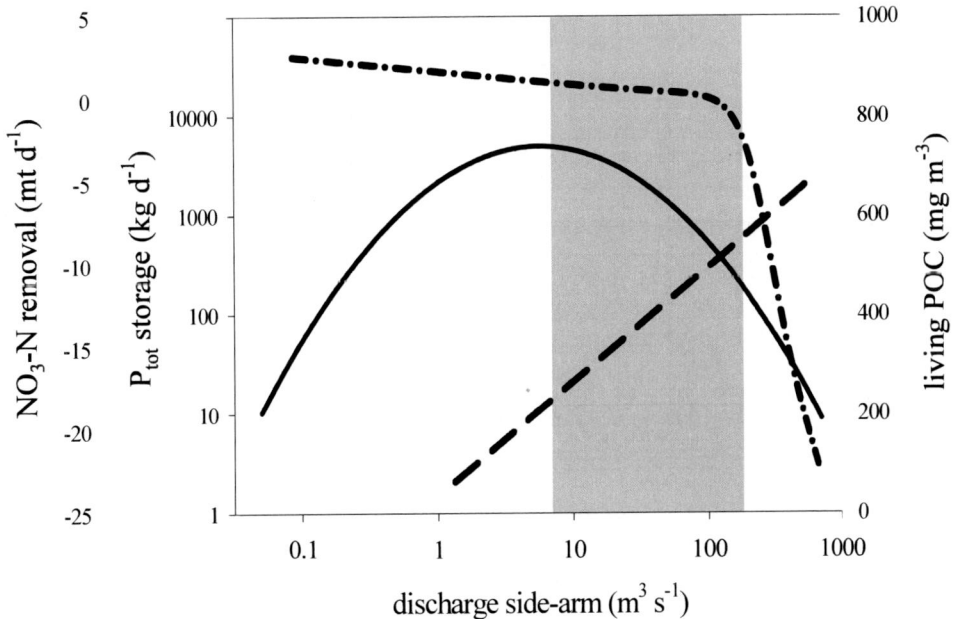

Figure 5. Significant relationships between side-arm discharge and total phosphorus (Ptot) – broken line, nitrate – broken-dotted line and living particulate organic matter (POC) – solid line. The living POC resembles the total planktonic carbon (for details see Hein et al. 2003). The data are from the years 1997 to 2000. The phosphorus relationship is presented in Hein et al. (2005), the living POC is from Hein et al. (2003). The nitrate removal was calculated on the basis of nitrate input and removal efficiencies (loss to atmosphere). The significance of the regressions is the following: for Ptot r^2= 0.74, n=87, p<0.01; for nitrate r^2= 0.69, n=55, p<0.01 and for living POC r^2= 0.21, n=34, p<0.01. The shaded area marks the increased discharge due to the rehabilitation measures. After Coops et al. 2006.

92   Thomas Hein, Elisabeth Bondar-Kunze, Friedrich Schiemer et al.

The difference in phosphorus retention and nitrate removal is striking for the years 2002 (a pronounced wet year) and 2003 (a dry year) (Fig. 6). In 2002 high phosphorus retention was imputed to 2 large flood events. In 2003 maximum nitrate removal rates coincided with a pronounced low water period in spring/summer. However, the variability between maximum and minimum removal and retention rates were quite different for phosphorus and nitrate, the P retention varied by an order of magnitude, while nitrate removal changed by a factor of 2.

Hence, any evaluation of rehabilitation measures needs to consider the spatial dynamics, the hot moments (McClain et al. 2003), and to estimate the effects on the side-arm. Considering the effect of the respective side-arm restoration on the matter dynamics in the whole river corridor, the hydrological exchange limited the overall significance of the increased transformation capacity, as at mean water only 1% of the discharge passed the side-arm system.

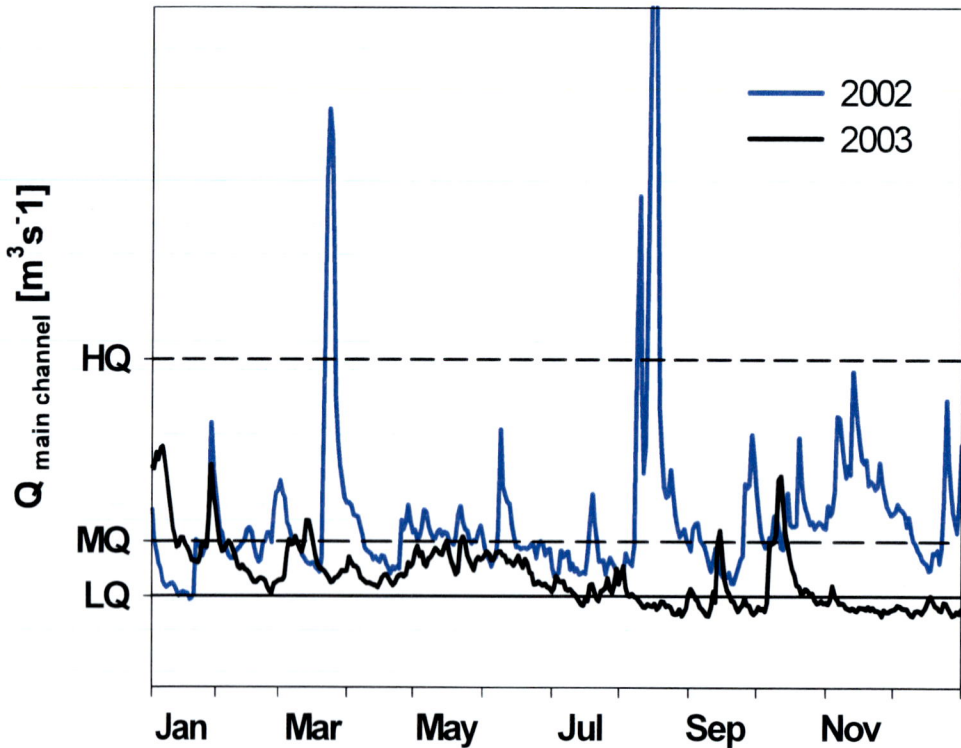

Figure 6. Comparison of discharge in the main channel of the Danube River. The years 2002 (in blue) and 2003 are presented, based on daily water level recordings.

Rehabilitation and restoration concepts for large river-floodplain systems require a profound insight into ecological functioning. The coupling of hydrology and ecological processes can play an important role in understanding large-scale biogeochemical processes and using ecosystem services for more effective river management (compare to the 3 principles). In urban and industrial areas, more natural exchange processes with retention areas can support other engineering-based solutions to achieve the required water quality goals (McClain 2002). In large regulated and /or straightened river systems, restoration and rehabilitation efforts should favor retention areas at different spatial scales, as shown for the Danube in Austria. In-channel structures functioning as storage zones are key elements during lower water tables and can substantially support the riverine food web in regulated rivers. With increasing flow, the availability of these in-channel structures declined due to the embankment, pointing to the need for other elements integrated in the riverine flow.

The presented side-arm reopening in the Austrian Danube provides an example of rehabilitation on the reach scale: rare lotic elements of the former braided reach were reintroduced during medium to high flows. Within the side-arm a more natural mass balance between storage, transformation and export of nutrients and organic matter was established based on increased hydrological exchange in concert with increased retention time and available area in the side-arm system (Hein et al. 2003). Increased nutrient transformation and retention also represent a socio-economically important, ecosystem service (Gren et al. 1995). During floods the loss of inundation area, combined with the lateral link to terrestrial components of the riverine landscape (Ward et al. 2002), can be expected to reduce the retention capacity compared with pre-regulation conditions. Hence, the rehabilitation of retention areas along the course of the river can affect local biogeochemical processes as well as enhance overall transformation capacity.

The nutrient retention function is based on specific processes. Nitrate removal is known to be closely linked to denitrification, thus, the environmental control and its changes are important to understand the capacity of nitrogen retention. With a change in the hydrological connectivity following reconnection, the nutrients reaching the specific area are transformed throughout the floodplain. As floodplain dynamics change, the nutrient pools of the floodplain change.

Following reconnection, the system dynamics of the Orth side arm changed significantly. Large surface water inflow from the Danube River brought large amounts of nitrate into the connected areas. However, as shown in previous studies (Johnson et al. 2007; Richardson 2004), nitrate concentrations reaching the backwaters of the floodplain are lower than those closer to the source river. The carbon dynamics also changed along this gradient. Areas closer to the river (highly dynamic) presented lower macrophytic density and decreasing amount of organic material in the sediment. In the more disconnected backwaters of the floodplain, denitrification was higher than in the highly dynamic and connected sites (Fig 7). Potential denitrification rates (modified from Yoshinari & Knowles 1976) have been measured for the Lobau and Orth floodplains in sites throughout the floodplains. The increased retention time plus the availability of labile carbon sources created more favourable conditions in more disconnected backwaters than in the carbon limited more connected areas. The controlling factors, caused by either carbon or nitrogen limitation within the floodplain controlled the potential for denitrification. With increasing water retention, the ratio of complete (N_2 emission) to incomplete (N_2O emission) denitrification increased, resulting in more N_2 production in the backwaters. However, within areas of similar connectivity, the presence of litter fall influenced the denitrification potential of the area, demonstrating the importance of the carbon source. Previous studies relating denitrification potential to substrate and carbon source also report a similar trend (Arango et al. 2007).

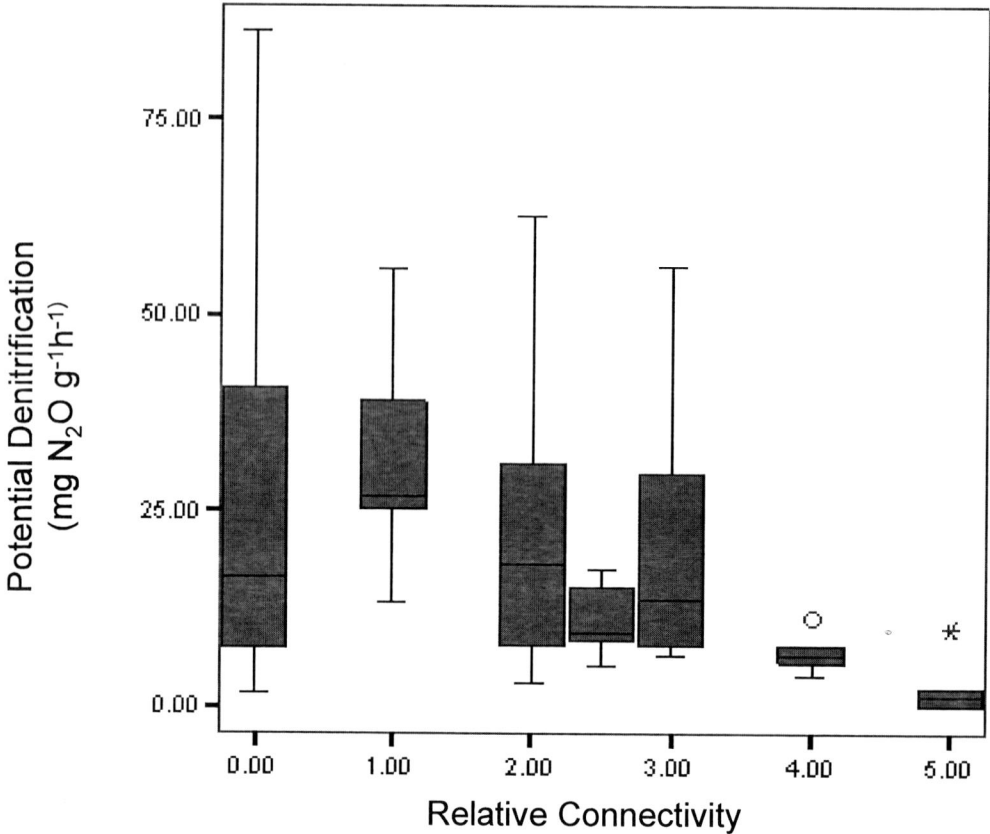

Figure 7. Box plot showing potential denitrification rates measured using the static core acetylene inhibition method (modified from Yoshinari & Knowles 1976). Rates are presented as mg N-N_2O per gram of dry sediment per hour. The whiskers show the highest and lowest measured rates with the median shown as the dark line inside the boxes. The boxes represent 50% of all measured rate with the top and bottom quartiles represented by the lines extending from the boxes. Sites were measured throughout the Orth and Lobau floodplains during the vegetation period in 2006 & 2007 (n = 87) at different relative connectivity levels using sediment slurries. The closer the site is to the main channel of the Danube, the higher the relative connectivity. Outliers are marked with a star and open circle.

THE IMPORTANCE OF ALGAL PRODUCTION
FOR ORGANIC MATTER DYNAMICS

The scientific debate about the role of allochthonous and autochthonous carbon sources in fueling and structuring the aquatic food webs has promoted research on characterizing and tracking the source of organic matter and estimating its utilization over the last 10 to 15 years (Thorp et al. 2006). Averaged over the year and river network, autochthonous autotrophy apparently provides >50% of the energy supporting metazoan production in channel sites and retention areas (Bunn et al. 2003). In contrast to these findings, another recent study pointed out the fact that fluvial networks degraded higher amounts of terrestrial carbon than originally thought (Battin et al. 2008). Exceptions will definitely occur where autotrophs are limited by shading or low light levels in the water column.

The trophic basis of production is expected to shift from a relative emphasis from benthic to pelagic autotrophs from headwaters to mouth. This shift is also influenced by the nature of local factors. The river continuum concept (Vannote et al. 1980) stipulated that the primary food sources along the longitudinal dimension of the river network were: (a) allochthonous organic matter (principally riparian leaves) in headwater streams with a heavy canopy cover; (b) benthic autotrophs in shallow mid-order rivers; and (c) fine particulate organic matter (FPOM) in large rivers that was derived from terrestrial organic matter via leakage from upstream food webs. The key question in ecological research raised at this time was: how different OM sources control varying components of aquatic food webs. Finlay (2001) showed that consumer $\delta^{13}C$ values in temperate headwaters and medium-sized rivers were more strongly related to algal than to terrestrial $\delta^{13}C$ values. A clear transition from terrestrial to algal carbon sources with increasing stream order was found which was linked to decreasing canopies.

The source of OM fueling large river food webs has been more controversial. A popular view (e.g. Sedell et al. 1989); and the flood pulse concept of (Junk et al. 1989) was that terrestrial detritus and aquatic macrophytes on submerged floodplains (rather than FPOM from upstream) were the predominant nutrient sources in floodplain rivers. Other studies have concluded from stable isotope data that the major annual energy source supporting overall metazoan production of most constricted and floodplain rivers was autochthonous primary production (Hamilton et al. 1992) and constitute major parts to the POM pool in large rivers as shown for the Upper Mississippi during summer (Delong & Thorp 2005). This production is supposed to enter food webs via algal-grazer and decomposer pathways (Thorp & Delong 2002). However, a decomposer (microbially mediated) food pathway may process most of the transported, allochthonous and autochthonous carbon and contribute substantially to the heterotrophic state (P/R < 1) of many large rivers (Thorp & Delong 2002). A weak coupling between microbial and metazoan production has also been noted by (Lewis et al. 2001).

Recent analyses of seston composition using stable isotope and C:N ratios demonstrated that living and detritic autochthonous matter, primarily phytoplankton, is a major constituent of transported POM. The predominance of autochthonous organic matter in seston has been shown for the Mississippi, Colorado, Rio Grande, and Columbia Rivers (Kendall et al. 2001), for a floodplain reach of the Danube River (Hein et al. 2003). Although seston composition varies among seasons and rivers, the predominance of autochthonous organic matter in seston extends for much of the year. The importance of allochthonous carbon to metazoans as a whole has been challenged recently for tropical and temperate floodplain rivers (e.g. Forsberg et al. 1993; Hamilton et al. 1992; Hamilton & Lewis 1992; Lewis et al. 2001). These studies support the dominant role of grazer and detritic consumption of algae in floodplain food webs.

Aspetsberger et al. (2002) and Hein et al. (2003) contributed to this debate by analyzing on a) how organic matter could be characterized by bulk parameters such as elemental ratios and stable isotopes and on b) how the organic matter dynamics were related to discharge conditions in dynamic side-arms. In both papers the authors demonstrated that algal and terrestrial sources were characterized by different isotope signatures and elemental ratios, and that these varied with different discharge conditions. If terrestrial material dominated during floods, organic matters of aquatic origin were predominant in the side arms during low to medium discharge conditions following floods. The primary production of algae was stimulated by the introduced nutrients during flooding, as proposed in the formulation of the third principle (see introduction). The depleted stable isotope values were in the range of various aquatic primary producers measured in the floodplain area (Hein, unpublished data). The enriched values found for carbon isotope signatures of transported POM during flooding agreed with the findings of (Füreder et al. 2003), who showed rather enriched values in streams of the Alpine region.

The differences in the nitrogen isotope signature of POM were related to the bacterial secondary production rates in the side-arm (Aspetsberger et al. 2002). The high supply of riverine nitrate, together with lower dissolved organic matter availability for microbes entailed a lower nitrogen isotope fractionation during high flows. During medium to low connectivity, however, mainly microbial processes lead to increased nitrogen isotope fractionation (Clement et al. 2003). Using the plankton contribution to POM and the carbon isotope signature, a carbon budget for the Regelsbrunn side-arm clearly demonstrated the transformation capacity during medium flows. This underlined the relevance of plankton sources as a subsidy for main channel conditions (Hein et al. 2003; Preiner et al. 2008). When considering the contribution of floodplain carbon to the entire riverine landscape carbon budget, two aspects must be considered even when surface connectivity events are infrequent, or rare (Junk & Wantzen 2004):

1. Flow conditions vary considerably between different river-floodplain systems. If geomorphology limits the exchange between river and floodplain, then the contribution of the floodplain carbon to the main stem carbon budget can be lower than expected (Lewis et al. 2001).
2. Mobile organisms such as fish actively seek floodplain carbon in mass migrations as soon as flooding begins (e.g. Wantzen et al. 2002). When small floodplain fish migrate back into the main channel during the falling limb of the hydrograph, they are heavily preyed upon by riverine predators (Wantzen et al. 2002).

Therefore, floodplain carbon can contribute significantly to river food webs even without a frequent surface water exchange; the hydrological connectivity characterized by the exchange during floods is the key factor.

Based on these findings the primary production rates of planktonic and benthic communities were investigated, with emphasis on situations during low flow conditions with increasing residence times (in our case study water ages). The production rates were used to find a link between the observed biomass development and the physiological background of the algae and demonstrate the effectiveness of these side-arm systems as proposed in the third principle of hydromorphological dynamics. The potential of algal productivity was estimated for a wide range of discharge conditions in connected side-arms. While planktonic production dominated at medium discharge conditions, benthic contribution was the decisive source during low water periods. These findings match quite well with conceptual models on the control of primary production in side-arms by hydrology and morphology (Hein et al. 2005).

Based on these findings we developed a conceptual scheme showing how connectivity patterns controlled the dominance of different aquatic primary producers in braided river systems (Fig. 8a). With increasing connectivity we expected a shift from macrophyte-dominated to benthic algae dominated systems, and with high surface water connectivity to planktonic algae. These shifts were mainly controlled by the direct effects of turbulence and the indirect effects of nutrient supply and light availability in the side-arms. In contrast to the common view about how trophic state influenced different groups of primary producers, we expected that primarily hydrology determines the distribution of all groups in dynamically connected retention areas (Fig. 8a+b). In isolated subsystems, where terrestrialization processes prevailed, productivity was controlled by nutrient availability and dominated by emergent macrophytes (compare Fig. 8b, after Brönmark & Hansson 2005). Thus, based on the frequency of surface water connections, namely flow and flood pulses, different groups of primary producers develop with a different resilience to these changes.

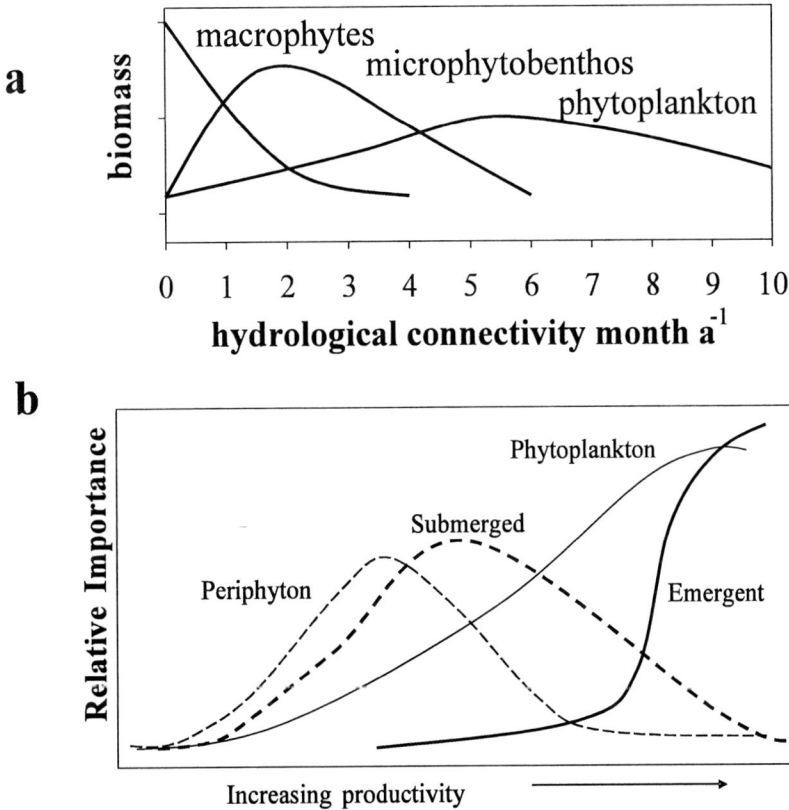

Figure 8. a) Scheme of the expected biomass distribution of selected aquatic primary producers in the Danube floodplains in relation to the hydrological connectivity (modified after Schiemer et al. 2006) b) Conceptual scheme of aquatic primary production in lakes and ponds after Brönmark & Hansson (2005).

This conceptual scheme needs further verification at two levels: i) at the individual side-arm scale (reach scale) the temporal and seasonal variability will be a major issue and ii) at the landscape scale (distribution of different connected side-arms) the summarizing effect on the carbon budget is crucial for an overall understanding of ecological restoration and is clearly linked to the proposed principles of hydromorphological dynamics. The latter level implies to question at which "time windows" transport vs. transformation processes regulate the carbon budget of riverine landscapes and how different groups of aquatic primary producers interact and are affected by restoration measures.

THE HYDROLOGICAL CONTROL OF PLANKTONIC PROCESSES IN SIDE-ARMS

The effect of hydrological connectivity on planktonic processes – phytoplankton and zooplankton biomass development, species diversity and their interaction (grazing of zooplankton on phytoplankton) was investigated and demonstrate at the level of food web interactions and species numbers the effects of the 3 principles of hydromorphological dynamics.

While the seasonal aspect of plankton development in temperate lakes has been extensively described (see for a summary in the PEG (Plankton Ecology Group) model), plankton succession and the interaction between zoo- and phytoplankton have traditionally been thought to be of minor importance in fluvial systems, where short retention times and frequent disturbance events dominated. Ongoing research during the last decade pointed to considerable plankton development in rivers, especially in large rivers and their retention areas (Dokulil & Teubner 2003).

In the River Meuse for instance, Descy & Gosselain (1994) found that phytoplankton development mainly originated from retention areas. In the Danube River, Reckendorfer et al. (1999) demonstrated that in-channel structures (shallow areas along the boundary of the channel) explained most of the zooplankton development found in downstream sections of the main channel. Different modeling approaches demonstrated that physical factors were the major controls of phytoplankton development in temperate rivers at large scales (Everbecq et al. 2001; Sellers & Buckaveckas 2003). Accordingly, the key predictors for phytoplankton biomass in large rivers are discharge and residence time, light availability (explained by mean depth and turbidity), temperature and nutrient levels to a lesser extent. Nonetheless, these models show that during certain periods, biotic interactions can become important. Recent results on the Rhine River also demonstrated that biotic interactions structured and impacted the microbial components of riverine food webs (Weitere et al. 2005). Among grazers the unicellular forms, especially heterotrophic flagellates, dominated (Weitere & Arndt 2002b; Weitere et al. 2005). Yet, they appeared to exert a moderate grazing pressure on phytoplankton communities.

The environmental control of plankton communities in floodplains is mainly related to the hydrological conditions. In the Regelsbrunn system, Baranyi et al. (2002) and Riedler et al. (2006) investigated how hydrological exchange affected zoo- and phytoplankton communities in terms of biomass development and species number. Phytoplankton biomass peaked at medium water ages in the side-arm of Regelsbrunn, coinciding with the maximum number of species observed. Here, a clear link between biomass and species diversity was found. This result supported a general ecosystem behavior in which maximum diversity was expected at biomass peaks. For metazoan zooplankton, a succession and interaction between rotifers and crustaceans was found (Baranyi et al. 2002). While rotifers established within a few days after high flows, the development of crustaceans took longer; this ultimately led to higher grazing rates on phytoplankton, explaining their decline (Keckeis et al. 2003; Riedler et al. 2006). Both papers showed the hydrological control of plankton species in a dynamic environment; Keckeis et al. (2003), even demonstrated that there was a clear shift between phases of abiotic and biotic control in plankton food webs of connected side-arms. Combining the results from the river-sidearm system of the Danube published in Keckeis et al. (2003)

with the protist components (Schiemer & Hein 2007) clearly demonstrated that microzooplankton presented different production patterns with water age. With increasing water age, the biomass of heterotrophic flagellates and ciliates increased at a moderate rate (Fig. 9). Both groups were less stimulated by retention compared to the fast initial bloom of phytoplankton.

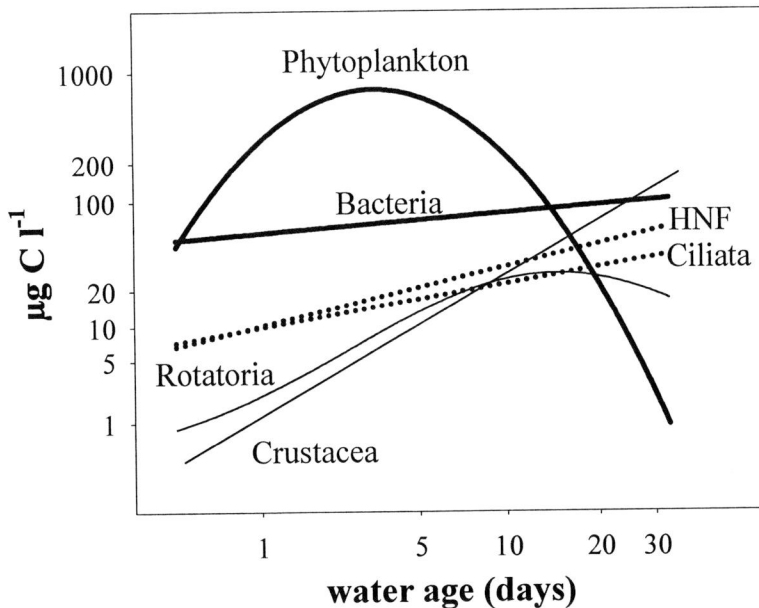

Figure 9. Biomass accumulation of plankton in side-arms of the Danube River following a flood pulse (after Schiemer et al. 2006). The x-axis gives water age in days in a logarithmic scale. The biomass of all functional groups of plankton is expressed as mg carbon per litre. The biomass values are grouped as phytoplankton, bacteria,, ciliates and heterotrophic nanoflagellates (HNF) and metazooplankton. The graph is based on a large set of partially unpublished data and published data (see Hein et al. 2003, Hein et al. 2005)

Microbial processes and accompanying unicellular grazers appeared to be less controlled by hydrological connectivity. In contrast to the heterotrophic flagellates and ciliates, the micro-metazoan plankton, especially rotifers, exhibited strong population growth at an early stage after a flood-pulse, followed by a distinct increase of small cladocerans (Bosmina). Under these conditions the microzooplankton exerted a strong predation pressure on the phytoplankton communities (Keckeis et al. 2003). This sequence can be explained by differences in respective generation times (Reckendorfer et al. 1999, Baranyi et al. 2002).

Summing up our present state of knowledge for restored side-arms, it appears that the pelagic food chain is primarily under hydrologic control. With increasing residence time, a change from dominating autotrophic to heterotrophic conditions and from physical to biotic control was found. For zooplankton there was a distinct sequence from unicellular to metazoan grazers. Higher biomass of crustaceans as efficient grazers was responsible for a decline of phytoplankton at longer retention phases.

Interactions may occur not only within the water column itself. There is growing evidence that, within large rivers, the coupling between the plankton community and benthic filter feeders (e.g. bivalves and amphipods) can be specifically important in slackwater areas; these areas have an enhanced biological activity, and the interaction influences the structure and function of "potamoplankton" (Thorp & Casper 2003). Another aspects is that overall retention capacity in large rivers is also influenced by microbial processes in the hyporheic zone (Fischer et al. 2005).

FUTURE PERSPECTIVES FOR ECOLOGICAL RESTORATION OF LARGE RIVERS

Future research has to strengthen the link between basic ecological principles and changing ecosystem conditions, and by that provide the scientific basis for restoration measures. Based on the current research activities, two aspects present promising future research avenues:

- A more elaborated, hierarchically structured understanding of processes across-scales in riverine networks including the extent of the river corridor
- Advanced complexity research by an integrated interdisciplinary indicator set that couples geomorphological, ecological and social sciences in ecological restoration

Lotic ecosystems are excellent systems for studying general ecological phenomena such as disturbance, succession, "criticality" and ecological stoichiometry in relation to restoration. The role of disturbances as a basic ingredient of floodplains and whole river corridors has been extensively discussed (Fisher et al. 1998; Sparks & Spink 1998; Townsend et al. 1997; Ward & Tockner 2001; Wooton et al. 1996). Disturbances, both natural and human induced, lead to changes in habitat turnover and interaction of various landscape elements (Ward et al. 2002). These successional patterns can completely change, as shown for habitat age distribution in a floodplain and associated vegetation patterns (Hohensinner et al. 2007). The research field of ecological stoichiometry, incorporating information on nutrient pools and trophic dynamics, has wide implications for food web dynamics under changing environmental conditions (Elser & Urabe 1999) and this understanding of ecosystem functions needs to be incorporated in applied issues related to ecosystem management. In rivers, frequent imbalances in C:N:P ratios that affect primary producers are expected to have far-reaching consequences for herbivores as well as for the microbial community (Woodward & Hildrew 2002). The basis for the temporally and spatially driven stoichiometric imbalances is the interaction with the principles of hydromorphological dynamics leading to the mosaic of subsystems with different nutrient and organic matter conditions and the hydrological exchange processes controlling light and nutrient availability. In particular, the spatial distribution of areas with increased water residence times, their sensitivity to disturbance events, and their rate of recovery after such disturbance events are important in interpreting the variability in elemental stoichiometry. These large-scale changes also influence the community structure and biotic interactions in rivers' floodplains.

Scaling is one of the major issues in ecology, and the link between scales for environmental and biotic processes is a key to understanding the reaction of ecosystems to human interventions (Thorp et al. 2006). Recently, the integrated view of river networks and landscape interactions (e.g. Benda et al. 2004; Gomi et al. 2002; Montgomery 1999) provided a conceptual basis at the catchment scale. Other concepts have examined processes operating at mid to small scales, using hierarchical habitat templates (Frissell et al. 1986; Poff 1997), flow regime (Lytle & Poff 2004; Thoms & Parsons 2003), inshore retention (Schiemer et al. 2001; Thoms et al. 2005), patch dynamics (Pringle et al. 1988; Townsend 1989) and river discontinuities (Poole 2002). Between different spatial scales, trans-scale processes (sensu Poole 2002) impact the ecosystems' structure, e.g., ecological disturbances (flood, fire, etc.) acting differently at the catchment scale and influencing the internal structure of the ecosystem at smaller scales. Today, the concept of bottom-up and top-down trans-scale linkages provides a foundation for understanding the geomorphologic dynamics and interactions with biological communities (Poole 2002). In order to develop a better foundation for understanding these large-scale processes and patterns, a "landscape ecology" approach is required (Turner et al. 1989) to link spatially explicit physical features described in the principles of hydromorphological dynamics to biota. The general theory in landscape ecology argues that heterogeneous spatial patterns matter, because they set the context for ecological processes such as fluxes of organisms, materials, and energy among landscape elements (Pickett & Cadenasso 1995; Wiens 1999). An important aspect of retention areas in riverine landscapes is how altered water regimes affect the nutrient cycling at different scales (Hill et al. 2000; Pinay et al. 1995). The main challenge, however, is to evaluate the effects of these changes at larger scales, corresponding to the landscape scale and under the context of human alterations (Gergel et al. 2005). The diversity of landscape elements, their turnover and interactions, hydrogeomorphic processes including the role of large woody debris, biological aspects including biodiversity at the landscape level (aquatic and terrestrial species) and the implication for conservation and restoration need to be integrated in future research (Tockner et al. 2002).

The second important future issue deals with a more integrated scientific approach coupling human and ecosystem orientated aspects. The complexity of riverine landscapes makes an integrative approach of different disciplines a pre-requisite for ecosystem restoration (Diggelen et al. 2001; Palmer et al. 2005). As restoration is based on human derived priorities and expectations, targets and evaluation need to include ecosystem and societal descriptors and benchmarking (Pahl-Wostl 2006; Palmer et al. 2005). Future activities for successful and sustainable management approaches need to link human-ecosystem interactions at the local, e.g., impact of regulation, and the catchment scale, e.g., altered sediment transport processes due to upstream dams (Jungwirth et al. 2002; Wohl 2005) and the issue of reversibility to a pristine status. In that respect, large-scale studies must build upon processes that operate at the reach scale (i.e., main channel sections, tributaries, floodplain segments). A hydro-geomorphic approach, ecosystem modeling, a hierarchical structure of the ecosystem indicators and decision support systems will be key tools for solving these challenges and for identifying the nodes among several spatio-temporal scales (e.g. Reichert et al. 2007). Thus, the role of environmental factors interacting and changing the reaction on restoration measures is of fundamental importance for the evaluation of restoration measures and setting the appropriate targets, especially in regions of intense human utilization (Pfadenhauer 2001). In that respect, hydrological exchange patterns and morphological structures play a central role in floodplain restoration (Henry et al. 2002) and are also key issues for the further development at the landscape level (Hein et al. 2006).

ACKNOWLEDGMENTS

The research was funded by the FWF, the Austrian Science Fund, project # P19907 and P11720 and the Alluvial Zone National Park. A grant for NW was provided by the Austrian Committee of the International Association for Danube Research (AC-IAD). Hydrological data for the main channel were provided by the Austrian River Authority. We are grateful to H. Kraill for chemical analyses.

REFERENCES

Ahearn, D. S., J. H. Viers, et al. *Freshwater Biology* 2006 *51*: 1417–1433 Priming the Productivity Pump: Flood Pulse Driven Trends in Suspended Algal Biomass Distribution across a Restored Floodplain.

Alexander, R. B., R. A. Smith, et al. *Nature* 2000 *403* (6771): 758-761 Effect of Stream Channel Size on the Delivery of Nitrogen to the Gulf of Mexico.

Allan, J. D. *Stream Ecology: Structure and Function of Running Waters*. Kluwer Academic Publishers: Dordrecht, 1995.

Alvarez-Cobelas, M., D. G. Angeler, et al. *Environmental Pollution* 2008 *In Press, Corrected Proof*: Export of Nitrogen from Catchments: A Worldwide Analysis.

Arango, C. P., J. L. Tank, et al. *Freshwater Biology* 2007 *52* (7): 1210-1222 Benthic Organic Carbon Influences Denitrification in Streams with High Nitrate Concentration.

Arnell, N. W. *Global Environmental Change* 1999 *9*: 5-23. The Effect of Climate Change on Hydrological Regimes in Europe: A Continental Perspective.

Aspetsberger, F., F. Huber, et al. *Arch. Hydrobiol.* 2002 *156* (1): 23-42 Particulate Organic Matter Dynamics in a River Floodplain System: Impact of Hydrological Connectivity.

Baker, M. A. and P. Vervier. *Freshwater Biology* 2004 *49* (2): 181-190 Hydrological Variability, Organic Matter Supply and Denitrification in the Garonne River Ecosystem.

Balser, T., K. McMahon, et al. *Plant and Soil* 2006 *289* (1): 59-70 Bridging the Gap between Micro - and Macro-Scale Perspectives on the Role of Microbial Communities in Global Change Ecology.

Baranyi, C., T. Hein, et al. *Freshwater Biology* 2002 *47*: 473-482 Zooplankton Biomass and Community Structure in a Danube River Floodplain System: Effects of Hydrology.

Basu, B. K. and F. R. Pick. *Journal of Plankton Research* 1997 *19* (2): 237-253 Phytoplankton and Zooplankton Development in a Lowland, Temperate River.

Battin, T. J., L. A. Kaplan, et al. *Nature Geoscience* 2008: 1-6 Biophysical Controls on Organic Carbon Fluxes in Fluvial Networks.

Behrendt, H. and D. Opitz. *Hydrobiologia* 2000 *410*: 111-122 Retention of Nutrients in River Systems: Dependence on Specific Runoff and Hydraulic Load.

Behrendt, H., J. Van Gils, et al. *Archiv Hydrobiologie Suppl* 2005 *158* (1-2 (Large Rivers 16/1-2)): Point and Diffuse Nutrient Emissions and Loads in the Transboundary Danube River Basin. Ii Long Term Changes.

Benda, L., N. L. Poff, et al. *Bioscience* 2004 *54* (5): 413-427 The Network Dynamics Hypothesis: How Channel Networks Structure Riverine Habitats.

Benke, A. C., I. Chaubey, et al. *Ecology* 2000 8 (10): 2730-2741 Flood Pulse Dynamics of an Unregulated River Floodplain in the Southeastern U.S. Coastal Plain.

Bennett, E. M., S. R. Carpenter, et al. *Bioscience* 2001 *51* (3): 227-234 Human Impact on Erodable Phosphorus and Eutrophication: A Global Perspective.

Brinson, M. M. *Wetlands* 1993 *13*: 65-74 Changes in the Functioning of Wetlands Along Environmental Gradients.

Brinson, M. M., H. D. Bradshaw, et al. *Significance of Floodplain Sediments in Nutrient Exchange between a Stream and Its Floodplain.*Dynamics of Lotic Ecosystems; Ann Arbor Science: Ann Arbor, Michigan, 1983 199-220.

Brönmark, C. and L.-A. Hansson *The Biology of Lakes and Ponds - Biology of Habitats*. Oxford University Press: New York, 2005.

Brunet, R. C., G. Pinay, et al. *Regulated Rivers: Research & Management* 1994 *9*: 55-63 Role of the Floodplain and Riparian Zone in Suspended Mater and Nitrogen Retention in the Adour River, South-West France.

Buijse, A. D., H. Coops, et al. *Freshwater Biology* 2002 *47* (4): 889-907 Restoration Strategies for River Floodplains Along Large Lowland Rivers in Europe.

Bunn, S. E., P. M. Davies, et al. *Freshwater Biology* 2003 *48* (4): 619-635 Sources of Organic Carbon Supporting the Food Web of an Arid Zone Floodplain River.

Canavan, R., A. Laverman, et al. *Hydrobiologia* 2007 *584* (1): 27-36 Modeling Nitrogen Cycling in a Coastal Fresh Water Sediment.

Carling, P. A. *Freshwater Biology* 1992 *28* (2): 273-284 The Nature of the Fluid Boundary-Layer and the Selection of Parameters for Benthic Ecology.

Clement, J. C., R. Holmes, et al. *J Appl Ecol* 2003 *40*: 1035-1048 Isotopic Invesitgation of Denitrification in a Riparian Ecosystem in Western France.

Coops, H., K. Tockner, et al. *Restoring Lateral Connections between Rivers and Floodplains: Lessons from Rehabilitation Projects*. Ecological studies, 190; Springer Press: 2006 15-32.

Dahm, C. N., K.W.Cummins, et al. *Restoration Ecology* 1995 *3*: 225-238 An Ecosystem View of the Restoration of the Kissimmee River.

Delong, M. D. and J. H. Thorp. *Oecologia* 2005 *147*: 76-85 Significance of Instream Autotrophs in Trophic Dynamics of the Upper Mississippi River.

Delong, M. D., J. H. Thorp, et al. *Regulated Rivers: Research and Management* 2001 *17*: 217-232 Responses of Consumers and Food Resources to a High Magnitude, Unpredicted Flood in the Upper Mississippi River Basin.

Descy, J.-P. and V. Gosselain. *Hydrobiologia* 1994 *289*: 139-155 Development and Ecological Importance of Phytoplankton in a Large Lowland River (River Meuse, Belgium).

Diggelen, R. v., A. P. Grootjans, et al. *Restoration Ecology* 2001 *9* (2): 115-118 Ecological Restoration: State of the Art or State of the Science?

Dokulil, M. T. and K. Teubner. *Hydrobiologia* 2003 *506-509* (1): 29-35 Eutrophication and Restoration of Shallow Lakes – the Concept of Stable Equilibria Revisited.

Elser, J. and J. Urabe. *Ecology* 1999 *80* (735-751): The Stoichiometry of Consumer-Driven Nutrient Recycling: Theory, Observation, and Consequences.

Everbecq, E., V. Gosselain, et al. *Water Research* 2001 *35* (4): 901-912 Potamon: A Dynamic Model for Predicting Phytoplankton Composition and Biomass in Lowland Rivers.

Finlay, J. C. *Ecology* 2001 *84*: 1052-1064 Stable Carbon Isotope Ratios of River Biota: Implications for Energy Flow in Lotic Food Webs.

Fischer, H., F. Kloep, et al. *Biogeochemistry* 2005 *76* (2): 349-371 A River's Liver - Microbial Processes within the Hyporheic Zone of a Large Lowland River.

Fischer, H., S. C. Wanner, et al. *Biochemistry* 2002 *61*: 37-55 Bacterial Abundance and Production in River Sediments as Related to the Biochemical Composition of Particulate Organic Matter (Pom).

Fisher, S. G., N. B. Grimm, et al. *Ecosystems* 1998 *1*: 19-34 Material Spiraling in Stream Corridors: A Telescoping Ecosystem Model.

Forsberg, B. R., C. Araujolima, et al. *Ecology* 1993 *74* (3): 643-652 Autotrophic Carbon-Sources for Fish of the Central Amazon.

Forshay, K. J. and E. H. Stanley. *Biogeochemistry* 2005 *75* (1): 43-64 Rapid Nitrate Loss and Denitrification in a Temperate River Floodplain.

Friedl, G. and A. Wüest. *Aquat. Sci.* 2002 *64*: 55-65 Disrupting Biogeochemical Cycles - Consequences of Damming.

Frissell, C. A., W. J. Liss, et al. *Environmental Management* 1986 *10*: 199-214 A Hierarchical Framework for Stream Habitat Classification: Viewing Streams in a Watershed Context.

Füreder, L., C. Welter, et al. *Internat. Rev. Hydrobiol.* 2003 *88* (3-4): 314-331 Dietary and Stable Isotope (Δ13c, Δ15n) Analayses in Alpine Strem Insects.

Gergel, S. E., S. R. Carpenter, et al. *Global Change Biology* 2005 *11* (8): 1352-1367 Do Dams and Levees Impact Nitrogen Cycling? Simulating the Effects of Flood Alterations on Floodplain Denitrification.

Gomi, T., R. C. Sidle, et al. *Bioscience* 2002 *52* (10): 905-916 Understanding Processes and Downstream Linkages of Headwater Systems.

Gren, I. M., K. Groth, et al. *Journal of Environmental Management* 1995 *45*: 333-345 Economic Values of Danube Floodplains.

Grubaugh, J. W. and R. V. Anderson. *Hydrobiologia* 1989 *174*: 235-244 Upper Mississippi River: Seasonal and Floodplain Forest Influences on Organic Matter Transport.

Hamilton, S. K., W. M. Lewis JR, et al. *Oecologia* 1992 *89*: 324-330 Energy Sources for Aquatic Animals in the Orinoco River Floodplain: Evidence from Stable Isotopes.

Hamilton, S. K. and W. M. Lewis JR. *Geochim. Cosmochim. Acta* 1992 *56*: 4237-4246 Stable Carbon and Nitrogen Isotopes in Algae and Detritus from the Orinoco River Floodplain, Venezuela.

Hedges, J. I., E. Mayorga, et al. *Limnol. Oceanogr.* 2000 *45* (7): 1449-1466 Organic Matter in Bolivian Tributaries of the Amazon River: A Comparison to the Lower Mainstream.

Hefting, M., B. Beltman, et al. *Environmental Pollution* 2006 *139* (1): 143-156 Water Quality Dynamics and Hydrology in Nitrate Loaded Riparian Zones in the Netherlands.

Heiler, G., T. Hein, et al. *Reg. Rivers Res. & Manag.* 1995 *11*: 351-361 Hydrological Connectivity and Flood Pulses as the Central Aspects for the Integrity of a River-Floodplain System.

Hein, T., C. Baranyi, et al. *Freshw. Biol.* 2003 *48*: 220-232 Allochthonous and Autochthonous Particulate Organic Matter in Floodplains of the River Danube: The Importance of Hydrological Connectivity.

Hein, T., C. Baranyi, et al. *Internat. Assoc. Danube Res.* 2004a *35*: 65–73 Hydrochemische Schlüsselparameter Zur Evaluierung Von Restaurationsmaßnahmen in Der Österreichischen Donau.

Hein, T., C. Baranyi, et al. *Science of the Total Environment* 2004b *328* (1-3): 207-218 The Impact of Surface Water Exchange on the Nutrient and Particle Dynamics in Side-Arms Along the River Danube, Austria.

Hein, T., G. Heiler, et al. *Regul. Rivers: Res. Mgmt.* 1999 *15*: 259-270 The Danube Restoration Project: Functional Aspects and Planktonic Productivity in the Floodplain System.

Hein, T., S. Preiner, et al. *Archiv für Hydrobiologie - Supplementbände* 2006 *158*: 557-576 Limnological Concepts as the Basis for Understanding River Networks:Perspectives for the Danube.

Hein, T., W. Reckendorfer, et al. *Archiv für Hydrobiologie Suppl: Large Rivers* 2005 *15* (1-4): 425-442 The Role of Slackwater Areas for Biogeochemical Processes in Rehabilitated River Corridors: Examples from the Danube.

Henry, C. P., C. Amoros, et al. *Ecological Engineering* 2002 *18* (5): 543-554 Restoration Ecology of Riverine Wetlands: A 5-Year Post-Operation Survey on the Rhone River, France.

Hill, A. R. *Nature* 1979 *281*: 291-292 Denitrification in the Nitrogen Budget of a River Ecosystem.

Hill, A. R., K. J. Devito, et al. *Biogeochemistry* 2000 *51*: 193-223 Subsurface Denitrification in a Forest Riparian Zone: Interactions between Hydrology and Supplies of Nitrate and Organic Carbon.

Hogan, D. M., T. E. Jordan, et al. *Wetlands* 2004 *24* (3): 573-585 Phosphorus Retention and Soil Organic Carbon in Restored and Natural Freshwater Wetlands.

Hohensinner, S., H. Habersack, et al. *River Research and Applications* 2004 *20* (1): 25-41 Reconstruction of the Characteristics of a Natural Alluvial River-Floodplain System and Hydromorphological Changes Following Human Modifications: The Danube River (1812-1991).

Hohensinner, S., H. Habersack, et al. *River Res. Applic.* 2007 *20*: 25-41 Reconstruction of the Characteristics of a Natural Alluvial River-Floodplain System and the Hydrogeomorphological Changes Following Human Modifications: The Danube River (1921-1991).

Hynes, H. B. N. *Verhandlungen der INternationalen Vereinigung für Theoretische und Angewandte Limnologie* 1975 *19*: 1-15 The Stream and Its Valley.

Johnson, T. E., J. N. McNair, et al. *Freshwater Biology* 2007 *52* (4): 680-695 Stream Ecosystem Responses to Spatially Variable Land Cover: An Empirically Based Model for Developing Riparian Restoration Strategies.

Jones, J. B. and R. M. Holmes. *Trends in Ecology and Evolution* 1996 *11*: 239-242 Surface-Subsurface Interactions in Stream Ecosystems.

Jungwirth, M., S. Muhar, et al. *Freshwater Biology* 2002 *47*: 867-887 Re-Establishing and Assessing Ecological Integrity in Riverine Landscapes.

Junk, W. J., P. B. Bayley, et al. (1989). The Flood Pulse Concept in River-Floodplain Systems. Internat. Large River Symp., Canad. Spec. Publ. of Fish. and Aquatic Sci.

Junk, W. J. and K. M. Wantzen. *Proceedings of the Second International Symposium on the Management of Large Rivers for Fisheries. R. L. Welcomme and T. Petr. Bang kok, FAO.* 2004 *2*: 117-149 The Flood Pulse Concept: New Aspects, Approaches, and Applications – an Update.

Keckeis, S., C. Baranyi, et al. *Journal of Plankton Research* 2003 *25* (3): 243-253 The Significance of Zooplankton Grazing in a Floodplain System of the River Danube.

Kendall, C., S. R. Silva, et al. *Hydrological Processes* 2001 *15* (7): 1301-1346 Carbon and Nitrogen Isotopic Compositions of Particulate Organic Matter in Four Large River Systems across the United States.

Lewis, W. M. J., S. K. Hamilton, et al. *J. N. Am. Benthol. Soc.* 2001 *20* (2): 241-254 Foodweb Analysis of the Orinoco Floodplain Based on Production Estimates and Stable Isotope Data.

Loaiciga, H. A., J. B. Valdes, et al. *Journal of Hydrology* 1996 *174*: 83-127. Global Warming and the Hydrological Cycle.

Lowrance, R., K. Sharpe, et al. *Journal of Soil and Water Conservation* 1986 *July-August*: 266-271 Long Term Sediment Deposition in the Riparian Zone of a Coastal Plain Watershed.

Lytle, D. A. and N. L. Poff. *Trends in Ecology and Evolution* 2004 *19*: 94-100 Adaptation to Natural Flow Regimes.

McClain, M. E. *The Application of Ecohydrological Principles for Better Water Resources Management in South America.*The Ecohydrology of South American Rivers and Wetlands; IAHS Press: Wallingford, 2002 193-206.

McClain, M. E., E. W. Boyer, et al. *Ecosystems* 2003 *6* (4): 301-312 Biogeochemical Hot Spots and Hot Moments at the Interface of Terrestrial and Aquatic Ecosystems.

Miller, J. R. and G. L. Russell. *Journal of Geophysical research* 1992 *97*: 2757-2764. The Impact of Global Warming on River Runoff.

Mitsch, W. J. and J. W. Day. *Ecological Modelling* 2004 *178* (1-2): 133-155 Thinking Big with Whole-Ecosystem Studies and Ecosystem Restoration--a Legacy of H.T. Odum.

Montgomery, D. R. *Journal of the American Water Resources Association* 1999 *35*: 397-410 Process Domains and the River Continuum Concept.

Naiman, R. J. and H. Décamps. *Annual Review of Ecology and Systematics* 1997 *28*: 621-658 The Ecology of Interfaces: Riparian Zones.

Pahl-Wostl, C. *Ecology and Society* 2006 *11* (1): The Importance of Social Learning in Restoring the Multifunctionality of Rivers and Floodplains.

Palmer, M. A., E. S. Bernhardt, et al. *J Appl Ecol* 2005 *42* (2): 208-217 Standards for Ecologically Successful River Restoration.

Pedroli, B., G. d. Blust, et al. *Landscape Ecology* 2002 *17* (1): 5-18 Setting Targets in Strategies for River Restoration.

Peterjohn, W. T. and D. L. Correll. *Ecology* 1984 *65*: 1466-1475 Nutrient Dynamics in an Agricultural Watershed: Observations on the Role of a Riparian Forest.

Petts, G. E., H. Moller, et al. *Historical Changes in Large Alluvial Rivers: Western Europe.*; Wiley, Chichester, 1989.

Pfadenhauer, J. *Restoration Ecology* 2001 *9* (2): 220-229 Some Remarks on the Socio-Cultural Background of Restoration Ecology.

Pickett, S. T. A. and M. L. Cadenasso. *Science* 1995 *269* (5222): 331-334 Landscape Ecology - Spatial Heterogeneity in Ecological-Systems.

Pinay, G., C. Arles, et al. *Arch. Hydrobiol.* 1989 *114*: 401-414 Topographic Influence on Carbon and Nitrogen Dynamics in Riverine Woods.

Pinay, G. and H. Decamps. *Regulated Rivers: Research and Management* 1988 *2*: 507-516 The Role of Riparian Woods in Regulating Nitrogen Fluxes between the Alluvial Aquifer and Surface Water: A Conceptual Model.

Pinay, G., B. Gumiero, et al. *Freshwater Biology* 2007 *52* (2): 252-266 Patterns of Denitrification Rates in European Alluvial Soils under Various Hydrological Regimes.

Pinay, G. and R. J. Naiman. *Archiv für Hydrobiologie* 1991 *123*: 187-205 Short Term Hydrologic Variations and Nitrogen Dynamics in Beaver Created Meadows.

Pinay, G., C. Ruffinoni, et al. *Biogeochemistry* 1995 *30* (1): 9-29 Nitrogen Cycling in 2 Riparian Forest Soils under Different Geomorphic Conditions.

Poff, N. L. *Journal of the North American Benthological Society* 1997 *16* (2): 391-409 Landscape Filters and Species Traits: Towards Mechanistic Understanding and Prediction in Stream Ecology.

Ponnamperuma, F. N. *Advances in Agronomy* 1972 *24*: 29-96 The Chemistry of Submerged Soils.

Poole, G. C. *Freshwater Biol* 2002 *47* (4): 641-660 Fluvial Landscape Ecology: Addressing Uniqueness within the River Discontinuum.

Preiner, S., I. Drozdowski, et al. *Freshwater Biology* 2008 *53* (2): 238-252 The Significance of Side-Arm Connectivity for Carbon Dynamics of the River Danube, Austria.

Pringle, C. M., R. J. Naiman, et al. *J. N. Am. Benthol. Soc.* 1988 *7*: 503-524 Patch Dynamics in Lotic Systems: The Stream as a Mosaic.

Pusch, M., D. Fiebig, et al. *Freshwater Biology* 1998 *40* (3): 453-495 The Role of Micro-Organisms in the Ecological Connectivity of Running Waters.

Raymond, P. A. and J. E. Bauer. *Nature* 2001 *409*: 497-500 Riverine Export of Aged Terrestrial Organic Matter to the North Atlantic Ocean.

Reckendorfer, W., H. Keckeis, et al. *Freshwater Biology* 1999 *41*: 583-591 Zooplankton Abundance in the River Danube, Austria: The Significance of Inshore Retention.

Reckendorfer, W., R. Schmalfuss, et al. *Archiv für Hydrobiologie Supplement* 2005 *155* (1-4): 613-630 The Integrated River Engineering Project for the Free-Flowing Danube in the Austrian Alluvial Zone National Park: Contradictory Goals and Mutual Solutions.

Reckendorfer, W. and A. Steel. *Abh. Zool.-Bot. Ges. Österreich* 2004 *34*: 19-30 Auswirkungen Der Hydrologischen Vernetzung Zwischen Fluss Und Au Auf Hydrologie, Morphologie Und Sedimente - Effects of Hydrological Connectivity on Hydrology, Morphology and Sediments.

Reichert, P., M. Borsuk, et al. *Environmental Modelling & Software* 2007 *22* (2): 188-201 Concepts of Decision Support for River Rehabilitation.

Ren, T., R. Roy, et al. *Applied Environmental Microbiology* 2000 *66*: 3891-3897 Production and Consumption of Nitric Oxide by Three Methanotrophic Bacteria.

Richards, G. R. *Journal of Climate* 1993 *6*: 546-559. Change in Global Temperature: A Statistical Analysis.

Richardson, W. S., E; Bartsch, L; Monroe, E; Cavanaugh, J; Vingum, L; Soballe, D. *Can. J. Fish. Aquat. Sci.* 2004 *61*: 1102-1112 Denitrification in the Upper Mississippi River: Rates, Controls, and Contribution to Nitrate Flux.

Riedler, P., C. Baranyi, et al. *Archiv für Hydrobiologie. Supplementband. Large rivers* 2006 *16* (4): 577-594 Abiotic and Biotic Control of Phytoplankton Development in Dynamic Side-Arms of the River Danube.

Roberts, J. and J. A. Ludwig. *J. Ecol.* 1991 *79* (1): 117-127 Riparian Vegetation Along Current-Exposure Gradients in Floodplain Wetlands of the River Murray, Australia.

Rosenberg, D. M., P. McCully, et al. *Bioscience* 2000 *50* (9): 746-751 Global-Scale Environmental Effects of Hydrological Alterations: Introduction.

Sadler, J. P., D. Bell, et al. *Biol. Conserv* 2004 *118*: 41-56 The Hydroecological Controls and Conversation Values of Beetles on Exposed Riverine Sediments in England and Wales.

Salo, J., R. Kalliola, et al. *Nature* 1986 *332*: 254-258 River Dynamics and the Diversity of Amazon Lowland Forest.

Schemel, L. E., T. R. Sommer, et al. *Hydrobiologia* 2004 *513* (1-3): 129-139 Hydrologic Variability, Water Chemistry, and Phytoplankton Biomass in a Large Floodplain of the Sacramento River, Ca, USA.

Schiemer, F., C. Baumgartner, et al. *Reg. Rivers Res. & Manag.* 1999 *15*: 231-244 Restoration of Floodplain Rivers: The Danube Restoration Project.

Schiemer F., T. Hein, et al. *Ecohydrology and Hydrobiology* 2006 *6*: 7-18 Hydrological control of system characteristics of floodplain lakes.

Schiemer, F. and T. Hein *The Ecological Significance of Hydraulic Retention Zones.* Hydroecology and Ecohydrology: Past, Present and Future; John Wiley & Sons, Ltd: West Sussex, 2007 405-417.

Schiemer, F., H. Keckeis, et al. *Arch. Hydrobiol. Suppl.* 2001 *135* (2-4): 509-516 The "Inshore Retention Concept" And Its Significance for Large Rivers.

Schmidt, W. F. *Science* 2001 *294* (5546): 1444-+ Ecology - a True-Blue Vision for the Danube.

Sedell, J. R., J. E. Richey, et al. (1989). *The Rcc: A Basis for Expected Ecosystem Behaviour of Very Large Rivers?* International Large River Symposium, Can. Sp. Pupl. Fish. Aquat. Sci.

Seitzinger, S. P., C. Kroeze, et al. *Estuaries* 2002a *25* (4b): 640-655 Global Patters of Dissolbed Inorganic and Particulate Nitrogen Inputs to Coastal Systems: Recent Conditions and Future Projections.

Seitzinger, S. P., R. V. Styles, et al. *Biogeochemistry* 2002b *V57-58* (1): 199 Nitrogen Retention in Rivers: Model Development and Application to Watersheds in the Northeastern U.S.A.

Seitzinger, S. P., R. V. Styles, et al. *Biogeochemistry* 2002c *57/58* (1): 199-237 Nitrogen Retention in Rivers: Model Development and Application to Watersheds in the Northeastern USA.

Sellers, T. and P. A. Buckaveckas. *Limnology and Oceanography* 2003 *48*: 1476-1487 Phytoplankton Production in a Large, Regulated River: A Modeling and Mass Balance Assessment.

Sparks, R. E. and A. Spink. *Reg. Rivers Res. & Manag.* 1998 *14*: 155-159 Disturbance, Succession and Ecosystem Processes in Rivers and Estuaries: Effects of Extreme Hydrologic Events.

Thoms, M. C., M. Southwell, et al. *Geomorphology* 2005: Floodplain–River Ecosystems: Fragmentation and Water Resources Development.

Thoms, M. C. and M. Parsons. *River research and applications* 2003 *19*: 443–457 Identifying Spatial and Temporal Patterns in the Hydrological Character of the Condamine-Ballone River, Australia, Using Multivariate Statistics.

Thorp, J. H. and A. F. Casper. *River Research and Applications* 2003 *19* (3): 265-279 Importance of Biotic Interactions in Large Rivers: An Experiment with Planktivorous Fish, Dreissenid Mussels and Zooplankton in the St Lawrence River.

Thorp, J. H. and A. D. Delong. *Oikos* 2002 *96* (3): 543-550 Dominance of Autochthonous Autotrophic Carbon in Food Webs of Heterotrophic Rivers.

Thorp, J. H., M. C. Thoms, et al. *River Research and Applications* 2006 *22* (2): 123-147 The Riverine Ecosystem Synthesis: Biocomplexity in River Networks across Space and Time.

Tockner, K., F. Malard, et al. *Limnol. Oceanogr.* 2002 *47*: 266-277 Nutrients and Organic Matter in a Glacial River- Floodplain System (Val Roseg, Switzerland).

Tockner, K., F. Malard, et al. *Hydrological Processes* 2000 *14*: 2861-2883 An Extension of the Flood Pulse Concept.

Tockner, K., A. Paetzold, et al. *Ecology of Braided Rivers*.Braided Rivers: Process, Deposits, Ecology and Management; Blackwell Publishing: Oxford, 2006 332-352.

Tockner, K., F. Schiemer, et al. *Regulated Rivers: Research & Management* 1999 *15* (1-3): 245-258 The Danube Restoration Project: Species Diversity Patterns across Connectivity Gradients in the Floodplain System.

Tockner, K., F. Schiemer, et al. *Aquatic Conserv: Mar. Freshw. Ecosyst.* 1998: Conservation by Restoration: The Managment Concept for a River-Floodplain System on the Danube River in Austria.

Tockner, K. and J. A. Stanford. *Environmental Conservation* 2002 *29* (3): 308-330 Riverine Flood Plains: Present State and Future Trends.

Townsend-Small, A., M. E. McClain, et al. *Limnology and Oceanography* 2005 *50* (2): 672-685 Contributions of Carbon and Nitrogen from the Andes Mountains to the Amazon River: Evidence from an Elevational Gradient of Soils, Plants, and River Material.

Townsend, C. R. *Journal of the North American Benthological Society* 1989 *8*: 36-50 The Patch Dynamics Concept of Stream Community Ecology.

Townsend, C. R., M. R. Scarsbrook, et al. *Limnology and Oceanography* 1997 *42*: 938-949 The Intermediate Disturbance Hypothesis, Refugia, and Biodiversity in Streams.

Triska, F. J., V. C. Kennedy, et al. *Ecology* 1989 *70*: 1893-1905 Retention and Transport of Nutrients in a Third-Order Stream in Northwestern California: Hyporheic Processes.

Turner, M. G., R. V. O'Neill, et al. *Landscape Ecology* 1989 *3* (3): 153-162 Effects of Changing Spatial Scale on the Analysis of Landscape Pattern.

Valett, H. m., J. A. Morrice, et al. *Limnology and Oceanography* 1996 *41*: 333-345 Parent Lithology, Surface-Groundwater Exchange and Nitrate Retention in Headwater Streams.

Van der Lee, G. E. M., H. O. Venterink, et al. *River Research and Applications* 2004 *20*: 315-325 Nutrient Retention in Floodplains of the Rhine Distributaries in the Netherlands.

Vannote, R. L., G. W. Minshall, et al. *Can. J. Fish. Aquat. Sci.* 1980 *37*: 130-137 The River Continuum Concept.

Wantzen, K. M., F. A. Machado, et al. *Aquatic Sciences* 2002 *64*: 239-251 Seasonal Isotopic Changes in Fish of the Pantanal Wetland, Brazil.

Ward, J. V. *J. N. Am. Benthol. Soc.* 1989 *8* (1): 2-8 The Four-Dimensional Nature of Lotic Ecosystems.

Ward, J. V. *Biol. cons.* 1998 *83*: 269-278 Riverine Landscapes: Biodiversity Patterns, Disturbance Regimes, and Aquatic Conservation.

Ward, J. V. and K. Tockner. *Freshw. Biol.* 2001 *46*: 807-819 Biodiversity: Towards a Unifying Theme for River Ecology.

Ward, J. V., K. Tockner, et al. *Freshw. Biol.* 2002 *47* (4): 517-539 Riverine Landscape Diversity.

Weilhoefer, C. L., Y. Pan, et al. *Wetlands* 2008 *28* (2): 472-486 The Effects of River Floodwaters on Floodplain Wetland Water Quality and Diatom Assemblages.

Weitere, M. and H. Arndt. *Freshwater Biol* 2002a *47* (8): 1437-1450 Top-Down Effects on Pelagic Heterotrophic Nanoflagellates (Hnf) in a Large River (River Rhine): Do Losses to the Benthos Play a Role?

Weitere, M. and H. Arndt. *Microb. Ecol.* 2002b *44*: 19-29 Water Discharge-Regulated Bacteria-Heterotrophic Nanoflagellate (Hnf) Interactions in the Water Column of the River Rhine.

Weitere, M., A. Scherwass, et al. *River Research and Applications* 2005 *21* (5): 535-549 Planktonic Food Web Structure and Potential Carbon Flow in the Lower River Rhine with a Focus on the Role of Protozoans.

Wiens, J. A. *The Science and Practice of Landscape Ecology.*Landscape Ecological Analysis: Issues and Applications; Springer: New York, 1999.

Wohl, E. *Ecology and Society* 2005 *10* (2): 1-16 Compromised Rivers: Understanding Historical Human Impacts on Rivers in the Context of Restoration.

Woodward, G. and A. G. Hildrew. *Freshwater Biol* 2002 *47* (4): 777-798 Food Web Structure in Riverine Landscapes.

Woodward, G., R. Thompson, et al. *Pattern and Process in Food Webs: Evidence from Running Waters.*Aquatic Food Webs-an Ecosystem Approach; Oxford University Press: Oxford, 2005 51.66.

Wooton, J. T., M. S. Parker, et al. *Science* 1996 *273*: 1558-1561 Effects of Disturbance on River Food Webs.

Wright, R. B., B. G. Lockaby, et al. *Soil Science Soc Am. J.* 2001 *65*: 1293-1302 Phosphorus Availability in an Artificially Flooded Southeastern Floodplain Forest Soil.

WWF. *Final report of the Danube Pollution Reduction Programme* 1999: Evaluation of Wetlands and Floodplains Areas in the Danube River Basin.

Yoshinari, T. and R. Knowles. *Biochemical and Biophysical Research Communications* 1976 *69*: 705-710 Acetylene Inhibtion of Nitrous Oxide Reduction by Denitrifying Bacteria.

Zweimüller, I., M. Zessner, et al. *Hydrological Processes* 2008 *22*: 1022-1036 Effects of Climate Change on Nitrate Loads in a Large River: The Austrian Danube as Example.

In: Ecological Restoration
Editors: George H. Pardue and Thomas K. Olvera

ISBN 978-1-60741-013-3
© 2009 Nova Science Publishers, Inc.

Chapter 4

Integrating Structural and Functional Metrics to Assess the Early Success of Salt Marsh Restoration

Keith Walters[1]

Department of Marine Science, Coastal Carolina University, Conway, SC, USA

Abstract

The restoration or creation of estuarine wetlands is predicated on two presumptions: marsh habitats have significant value that if lost will produce negative effects on coastal ecosystems, and engineered wetlands are equivalent to historical marsh systems. Few studies directly have documented wetland value, but a greater number have attempted to assess the equivalency of restored and existing marshes. A majority of the restoration studies focus on monitoring the convergence of certain metrics (e.g., plant density, resident species richness) between marshes. Monitoring structural changes over time can be labor intensive, often requires decades of data before observing convergence, and potentially are confounded by ecosystem resilience, modified constraints to community change, and the existence of alternative stable states for systems. Recent advocacy of an experimental approach to assessing marsh restoration progress may offer a number of advantages over strict monitoring. An experimental approach was applied to the study of early progress in a Southeastern US created salt marsh. A 6 ha pine upland along the North River, GA was graded to intertidal elevations and planted with marsh vegetation (primarily *Spartina alterniflora*) as part of a federally mandated mitigation project. A series of experiments were conducted to assess the functional similarity of the created and an existing reference marsh along the same tidal creek. Structural characteristics of the marshes were both similar (e.g., stem densities, particulate settlement) and different (e.g., plant nearest neighbor distances, sediment organic content), but would suggest that after 3-4 yrs. the restored marsh had not progressed to the point of resembling the existing marsh. However, experimental results suggested restored and existing marshes functionally are similar. The growth and survival of caged bivalves and marked snails typically resident in marshes either were not significantly different between marshes or

---

1 Correspondence to: Department of Marine Science, P.O. Box 261954, Coastal Carolina University, Conway, SC 29528, Phone: 843.349.2477, email: kwalt@coastal.edu

were greater within the restored marsh. Foraging crabs appeared to exhibit a preference for natural marsh sediments, but feeding experiments indicated similar effects on sediments from both marshes. The survival of enclosed or tethered mollusks was similar between marshes except for a single instance and suggested marshes provide the same refuge function. Application of an experimental approach led to the conclusion that the restored marsh had progressed in a relatively short time to exhibit many ecological functions existent in historical marshes. An experimental approach also should provide greater insight into the potential reasons for why created marshes may not progress to historical targets.

Key Phrases: salt marsh functional similarity, monitoring marsh restoration progress, mollusk shell and tissue growth strategies

Keywords: *Spartina alterniflora*, feeding preference, habitat restoration, refuge effect, enclosure experiments, fiddler crabs, *Geukensia demissa*, Southeastern marshes, free-range snails, tethering, success metrics

INTRODUCTION

The relatively recent emphasis on ecological restoration of salt marsh habitats either by creation of new or rehabilitation of existing marshes acknowledges both the value of marshes to nearshore environments (e.g., Constanza et al. 1997) and an ability to alleviate most or all of the negative effects of marsh loss (Dobson et al. 1997, Young et al. 2005). Active restoration of U.S. estuarine marshes dates from the 1970s (Seneca et al. 1985, Broome et al. 1986) and typically involves the transplantation of one or a few marsh plant species such as salt marsh cordgrass, *Spartina alterniflora*. Although qualified successes exist (Roman et al. 2002, Warren et al. 2002, Teal & Peterson 2005, Baustian & Turner 2006, Konisky et al. 2006), a consensus on the efficacy of past restoration efforts has not been reached (Matthews & Minello 1994, Craft et al. 1999, Zedler & Callaway 1999, Peterson & Lipcius 2003, Luken & Walters 2009).

The ability to document success of any restoration project first involves the identification of appropriate goals and metrics (e.g., Weinstein et al. 1997, Short et al. 2000, Luckenbach et al. 2005). Although increasing emphasis is placed on societal objectives (Weinstein et al. 2001, Higgs 2005), reestablishment of ecological function continues to remain a major goal for marsh restoration. Historically, re-creating the structural aspects of the marsh (*e.g.*; plant and animal biomass, density or species composition) was viewed as sufficient for the return of ecological function (Simenstad & Thom 1996, Piehler et al. 1998). Early studies of natural and restored marsh structure suggest that macrophyte biomass recovered quickly (Seneca et al. 1985, Broome et al. 1982, 1986, Broome 1990), but significant differences in sediment microbial nitrogen dynamics (Thompson et al. 1995, Currin et al. 1996), nutrient cycling (Craft et al. 1988, Langis et al. 1991, Craft et al 1991), benthic infaunal abundance and composition (Moy & Levin 1990, Sacco et al. 1994, Levin et al. 1996), nekton marsh utilization (Minello & Zimmerman 1992, Allen et al. 1994, Minello & Webb 1997), and avian nesting behavior (Zedler 1993) continued many years after project completion. Longer term and/or more recent studies indicate attributes of restored marsh structure can equal values from native marshes if factors including time since restoration are considered (Posey et al. 1997, Craft et al. 1999, Brady et al. 2002, Keer & Zedler 2002, Warren et al 2002, Able et al. 2004, Konisky et al. 2006). Unfortunately, the development of marsh structural equivalency over time represents only a first step towards functional equivalency and can perpetuate the myth of successful ecological restoration (Hilderbrand et al. 2005).

A frequent lack of adequate experimental designs and confounded analyses also has plagued restoration studies that attempt to document success (Chapman 1999, Grayson et al. 1999, Holl et al. 2003). One pervasive difficulty is a lack of replication at the appropriate treatment level. To logically infer restoration is the cause of similarities or differences between habitats traditionally would require assessing the variation across multiple restored sites (see Hurlbert 1984). Unfortunately, restoration efforts typically are "projects of opportunity" for reasons that include site availability and lack of financial resources. Restored sites rarely are replicated (but see Craft et al. 1999, Walters & Coen 2006). Alternative design and analysis approaches have been developed to avoid the replication issue and include, naming a few, Beyond BACI (Underwood 1994), Bayesian statistics (Dixon & Ellison 1996), information-theoretic models (Burnham & Anderson 1998), and likelihood methods (Hilborn & Mangel 1997).

Another alternative approach to identifying similarities and/or differences in ecological functions between natural and restored marshes is the use of targeted, manipulative experiments. As has been demonstrated many times at the population, community or ecosystem level the use of manipulative experiments is one of the most powerful approaches for identifying differences in ecological processes (e.g., Carpenter et al. 1996). The objective of this study was to conduct a series of manipulative experiments to determine if restored and natural marshes are similar in two critical ecological functions identified for salt marsh systems (Nixon 1980, Dame 1989): (1) a source of food for marsh fauna; (2) a refuge from predation. Specific questions addressed with manipulative experiments included: Is the growth of resident bivalve and gastropod fauna different between restored and natural marshes? Are decapod feeding rates different between restored and natural marshes? Are predation rates on and survival of resident marsh bivalves and gastropods different between restored and natural marshes? Along with the experimental investigations structural metrics including floral and faunal densities and sediment salinities and organic content were measured to determine if a connection between physical structure and ecological function could be established. Results from this study will identify whether differences exist in a number of ecological processes between natural and restored marshes and will provide insights into potential explanations for the differences in ecological functions.

MATERIALS AND METHODS

All studies were conducted within adjacent natural and restored Kamehameha salt marshes on the Naval Submarine Base at Kings Bay, GA, USA (81°32'W, 30°47'N). The Kamehameha marshes are located along the North River, a drowned tidal river with precipitation and run-off the only freshwater input. Extensive natural salt marshes predominated by *S. alterniflora* surround the North River from an origin on the Naval Base to a terminus in the St. Marys River. In spring 1995 ca. 6 ha of pine upland were logged and all logs, stumps, shrubs and brush removed. Matted roots and roots >5.1 cm in diameter were grubbed from the cleared area to a depth of 45.7 cm. A series of tidal channels were dug to connect the acreage to the adjoining North River and the cleared area was graded to establish typical intertidal elevations. Cleared intertidal areas between 12.7 and 17.8 cm in elevation were planted in November 1996 with 10.2 cm plugs of *S. alterniflora* placed 0.9 m on center.

An elevational gradient in plant morphology and composition typical of southeastern saltmarshes (Wiegert & Freeman 1990) existed within natural and, to an extent, restored Kamehameha marshes at the beginning of the study in May 1997. *Spartina* stands in low-marsh elevations directly along the tidal channels were taller (>1 m) compared to the shorter (<0.5 m) stands in high-marsh elevations closer to the upland areas. The less frequently flooded high-marsh areas bordering existing pine uplands were fringed by stands of *Juncus roemerianus*. In the restored marsh only, *J. roemerianus* stands were interspersed with *Typha* sp., *Scirpus* sp., *Aster* sp. and other less common saltmarsh plant species.

Marsh Characteristics

Various biotic and abiotic characteristics were measured within the natural and restored Kamehameha marshes during the 1997 to 1999 study. Live and standing-dead *S. alterniflora* stem and shoot densities were enumerated in May from 0.0625 m^2 quadrats (n = 10) haphazardly placed within mid-marsh (1997-99) and low- and high-marsh zones (1998-99). Live stem heights (1997) within quadrats and live and dead stem nearest neighbor distances (1998) also were measured within the mid-marsh. Mussel, *Geukensia demissa*, densities were determined from each quadrat in 1998 and 1999. Sediment was sampled within each marsh zone (high-marsh samples from within *J. roemerianus* stands) in 1997-98 using a 2.1 cm diameter corer (n = 5) to a depth of 2 cm. Individual cores immediately were frozen at -20 °C for later processing.

Sediment samples either were processed to determine pore water salinity and sediment grain size or organic content and pigment amounts. Pore water salinities were determined from refractometer readings of thawed sediment samples hydrated with 20 ml of distilled water, repeatedly mixed and allowed to settle over 48 h and adjusted for dilution. The sediment silt/clay fraction (<63 µm) was determined by measuring the dry mass of wet-sieved standard soil size fractions. Organic content, ash-free dry mass (AFDM), was calculated from 2.5 cm^3 homogenized subsamples as the difference between sediments dried at 60 °C and then ashed at 450 °C for >4 h. Pigments were extracted from homogenized 1 cm^3 sediment subsamples treated with 90% acetone (with $MgCO_3$), shaken vigorously, and refrigerated for 24+ hr in the dark. Refrigerated samples were centrifuged, the supernatant processed in a PerkinElmer Lambda™ 40 UV/Vis spectrophotometer, and chlorophyll *a* amounts calculated as in Parsons et al. (1984).

Growth and Feeding Experiments

The growth and/or foraging of 3 typical marsh resident species, the ribbed mussel *Geukensia demissa*, the saltmarsh periwinkle *Littoraria irrorata,* and the fiddler crab *Uca pugilator* were followed within natural and restored marshes in enclosure, mark-recapture, and mesocosm experiments. Marsh mussels are capable of filtering microalgae, bacteria and flagellates from the water column (Kemp et al. 1990, Kreeger & Newell 2001); marsh periwinkles feed on detritus, microalgae, fungi, and *S. alterniflora* (Barlöcher & Newell 1994, Silliman et al. 2003); fiddler crabs consume organic material from the sediments (Robertson & Newell 1982). The use of a filter-feeding mussel and substrate-feeding snails and crabs for study was designed to detect effects of potential differences between natural and restored pelagic and benthic food availability. If zones and hydrological conditions were similar between natural and restored marsh sites than the growth of organisms dependent on the water column for food (e.g., mussels) should be similar between marshes. However organisms dependent directly on the marsh for food (e.g., snails and crabs) might indicate potential differences that could be ascribed to functional dissimilarities between natural and restored marshes.

Mussel growth was studied in 1997-98 and 1998-99 experiments using methods similar to those of Bertness and Grosholz (1985). Medium-sized mussels (35 – 49 cm in 1997, 44 – 76 cm in 1998) were collected from mid-marsh zones within a common natural marsh along the North (1997) or St. Marys River (1998). Use of mussels collected from a marsh site a significant distance from the experimental marshes is similar to a common garden experiment (e.g., Gallagher et al. 1988) and has the advantage of minimizing possible confounding site effects on survival and growth. Also the 1997 absence of mussels in the restored marsh prevented use of a reciprocal transplantation approach. Shell length, umbo to opposite shell margin, of all mussels was measured and identifiable individuals placed in natural and restored marshes within 0.64 cm Vexar® mesh enclosures 10 – 15 cm in diameter and ≈50 cm tall. Enclosures were placed within holes excavated in the sediment between *S. alterniflora* culms. Each enclosure was filled with sediment to the level of the existing surface to allow mussels to bury (Lent 1969, Franz 1997). Enclosures were designed to minimize potential losses from predation. Measured mussels were placed haphazardly within enclosures (1997 = 1 ea., 1998 = 3 - 5 ea.) positioned within mid- and low-marsh zones (1997 = 10 elev^{-1}, 1998 = 15 clcv^{-1}). All enclosed mussels were collected ≈12 mo. after field placement and immediately frozen for later processing. Additional mussels were collected from the North and St. Marys' natural marsh sites at the start of each experiment and immediately frozen. The additional mussels were processed and data used to predict initial shell and ash-free tissue mass of experimental mussels. Frozen mussel shell length was measured in the lab, all visible tissue removed from shells, and both shell and tissue dried at 60 °C for >2 d prior to determining dry mass.

To assess possible differences in pelagic food availability settlement traps were deployed in June and August 1998 to estimate the amount of material transported through the water column to each mid- and low-marsh zone where mussels were enclosed. Individual traps (n = 10) with an aspect ratio >5 were buried with ca. 1 cm above the sediment surface to minimize accumulation of material horizontally transported along the benthic boundary layer. Deployments lasted from 3-4 d and all material collected by the end of the sampling period was frozen immediately at -20 °C and stored until processed. Stored settlement samples initially were thawed and filtered through pre-ashed, pre-weighed Whatman GF/F filters. All filters were dried at 60 °C for >2 d to determine total particulate dry mass and ashed at 450 °C for >4 h to calculate AFDM.

Separate *L. irrorata* mark-recapture growth experiments were conducted in 1997 and 1998 within natural and restored Kamehameha marshes. All snails (>500) initially were collected haphazardly from mid-marsh zones within a common site along the St. Marys River. The collection of snails from a common site had the same advantages as in the mussel enclosure experiments (see above) and was necessitated by a complete absence of naturally occurring *L. irrorata* in both natural and restored Kamehameha marshes. Exhaustive surveys of either Kamehameha marsh throughout the study did not find a single, naturally occurring *L. irrorata*, although other saltmarsh snail species (e.g., *Melampis bidentatus*) were common (Walters, pers. obs.). Collected snails either were frozen immediately for later processing to predict initial shell and tissue mass (>100) or individually numbered with bee tags (Chr. Graze, FRG) and shell length, tip of apex to edge of operculum opening, and wet mass after toweling dry measured. Numbered snails (50 ea.) were released at 4 haphazardly selected sites within both Kamehameha natural and restored marshes. Sites were located within >1 m height stands of *S. alterniflora* and widely separated, >20 m apart, along a total distance of ca. 300 m in either marsh.

Comprehensive efforts to recover all marked, free-ranging snails were conducted in August 1997 and 1998. The length and wet mass of all recaptured snails were measured before all individuals were frozen at -20 °C. Shells of recaptured snails in the 1997 experiment were cracked and all tissue separated from the shell. Cracked shell and tissue were dried at 60 °C for >2 d to determine dry mass.

In August 1998 feeding and preference experiments were conducted using only female *U. pugilator* placed into 500 ml containers with varying sediment treatments. Crabs were collected from a single intertidal marsh site 24 hrs before the start of the experiment and starved. On the day of the experiment sediments were collected intact to a depth of 2 cm from natural and restored marshes along tidal creeks where crabs were observed foraging at low tides. In the feeding experiment undisturbed natural or restored marsh sediments were placed within each container. Half of the containers with either marsh sediment (n = 5) were left as controls and one crab each was placed in the remaining half (n = 5). In the preference experiment containers were filled with ½ natural and ½ restored marsh sediments. Again half the containers (n = 5) were left as controls and one crab each was placed in the remaining half (n = 5). The ½ - ½ containers were designed to assess whether crabs were able to distinguish between natural and restored marsh sediments when offered the choice. Additional sediment samples (n = 60 ea.) were collected from the same tidal creek sites to determine initial differences between natural and restored marshes.

Both feeding and preference experiments ran for ≈48 hr. under natural light conditions. Crab escape was prevented with a mesh top and sediments were kept moist. At the termination of the experiments sediment to a depth of 2 cm was collected using a 10 cc syringe corer. A total of 12 cores were collected from each container; 6 cores from each sediment side in the preference experiment. Sediment cores and crabs were frozen immediately at -20 °C and returned to the lab for processing. Homogenized 2 cm^3 subsamples were dried at 60 °C for >2 d to determine sediment dry mass and ashed at 450 °C for >4 h to calculate AFDM.

Survival Experiments

Relative differences in the survival of natural and restored marsh mussels were determined in the mussel enclosure experiments and a June to August 1998 tethering experiment (Heck & Wilson 1987, Aronson & Heck 1995). Enclosure experiments measured the effects of differences in physical stress and/or bottom-up resource availability on mortality. Tethering results measured relative differences in top-down or refuge effects (Rozas & Odum 1988, Hovel et al. 2001, Bortolus et al. 2002). Individual mussels collected from the St. Marys mid-marsh site in June 1998 were attached to a 1-2 m length of monofilament line. Tethered mussels (n = 3) were tied to survey flags (n = 5) that were placed within natural and restored mid- and low-marsh sites. The survival of mussels was recorded throughout the first 5 d and finally in August 1998. Mussels that were no longer tethered and missing (e.g., not located within the immediate vicinity of identified mussels), present but with noticeable shell perforation or present with no living tissue were classified as dead. Although the efficacy of tethering techniques have been questioned (Peterson & Black 1994, Kneib & Scheele 2000 but see Aronson et al. 2001) the approach has been used effectively to provide comparative estimates of organism loss (Clark et al. 2003, Eggleston et al. 2005, Orth et al. 2007). For the relatively immobile mussels, tethering primarily aided locating buried individuals and did not appear to restrict normal behavior. Given the potential for artifacts in tethering experiments results are interpreted cautiously.

Natural and restored marsh differences in *L. irrorata* survival also were examined in two tethering experiments conducted between May and July 1998. Individual snails were attached to a 1-2 m length of monofilament line, tied to survey flags and placed in the field. In May 40 snails were placed within the low-marsh zone among *S. alterniflora* stems in both natural and restored marshes. In June 40 snails again were placed in the low-marsh zone of natural and restored marshes but 10 in each marsh were placed outside of the leading marsh edge on the adjacent mud flat. The survival of tethered snails was determined daily for 5 d and ca. 30 d later when each experiment was terminated. Individuals no longer tethered and missing (e.g., tethers cut and identifiable snails not within the immediate vicinity) and present but with no living tissue were classified as dead; individuals that became entangled with *S. alterniflora* and committed suicide (e.g., snails suspended above the sediment with no access to stems) were not considered to have suffered predation. Snail tethering results also are interpreted cautiously especially given the stem-climbing behavior of *L. irrorata* (Vaughn & Fisher 1988, Hovel et al. 2001) and the observed but infrequent tendency to suicide.

Statistical Analyses

Natural and restored marsh differences in abiotic and biotic parameters were analyzed, where possible, using a blocked, 2-factor ANOVA or MANOVA. Marsh zones were the blocking factors to account for typical effects of spatial variation (e.g., low- to high-marsh) while focusing on date and marsh effects. ANOVA was applied to univariate measures of marsh success (e.g., stem height, sediment salinity) and MANOVA was used for data where dependent variables had an obvious multivariate structure (e.g., live and dead stems and shoots). An alternative nested ANOVA design was applied to stem height data (stems nested within quadrats, marsh effects tested over quadrat within marsh mean square error) and a 2-

factor ANOVA was applied to nearest-neighbor distances (marsh and stem type). A lack of marsh treatment replicates common in restoration studies (see Introduction) reduced the generality of inferences that can be drawn from the results of monitored biotic and abiotic parameters. All data were evaluated and appropriately transformed if data violated critical ANOVA/MANOVA assumptions. MANOVA tests were evaluated based on the generalized F of Wilks' lambda distribution which provided greater power and immunity from violations of assumptions.

Mussel and snail growth and feeding experiments also were analyzed using parametric procedures and ANOVA tests. To determine the initial shell and tissue mass for experimental mussels and snails, dry mass by total length least square regressions were generated from a sacrificed, representative sample collected from within the common North or St. Marys River marshes. The growth or change in shell or tissue mass was calculated by subtraction of regression-estimated initial from observed final values. Results were analyzed by 2-factor ANOVA blocking for zone effects in the mussel experiments and site effects in the snail experiments. Cage effects on mussel growth were tested for in 1998 and, if significant, included in the model.

Crab feeding and preference experiments were analyzed either as a nested ANOVA (container nested within marsh and crab treatments) or a blocked MANOVA. Analyzing feeding preference experiments can be problematic (Peterson & Renaud 1989), but suggestions to include appropriate controls (Peterson & Renaud 1989) and use a multivariate approach (Roa 1992, Manly 1993, Lockwood 1998) were incorporated into the analysis of crab preference data. Natural and restored marsh AFDM, the non-independent feeding choice results, were the dependent variables, containers were the blocking factor and crab presence or absence the treatment effect analyzed in the MANOVA.

Hierarchical loglinear models were used to analyze count data for snail mark-recapture and mussel and snail tethering experiments. A model selection process based on the change in likelihood ratio χ^2 values was applied. All statistical tests were run on SAS 9.1 (Clark 2004) or SPSS 16.0 for Windows (Norušis 2008).

RESULTS

Marsh Characteristics

Live and dead *S. alterniflora* stem and shoot numbers varied appreciably among marshes and years but not marsh zones (Fig. 1). Densities were significantly different between natural and restored mid-marsh zones analyzed separately for 1997 (MANOVA, $F_{3, 16} = 12.17$, $p < 0.001$) and were affected by a significant year*marsh interaction (MANOVA, $F_{3, 112} = 34.03$, $p < 0.001$) but not a significant zone or block effect for 1998 to 1999 (MANOVA, $F_{6, 224} = 1.36$, $p > 0.05$). Low- and high-marsh live stem densities increased by ≈50% in natural and declined by ≈40% in restored marshes between 1998 and 1999 (Fig. 1). Rarely did restored exceed natural marsh stem densities, but in 1998 shoot densities were at least twice as great in all restored compared to natural marsh zones (Fig. 1). However, the increased shoot production did not result in greater numbers of restored marsh live stems the next year.

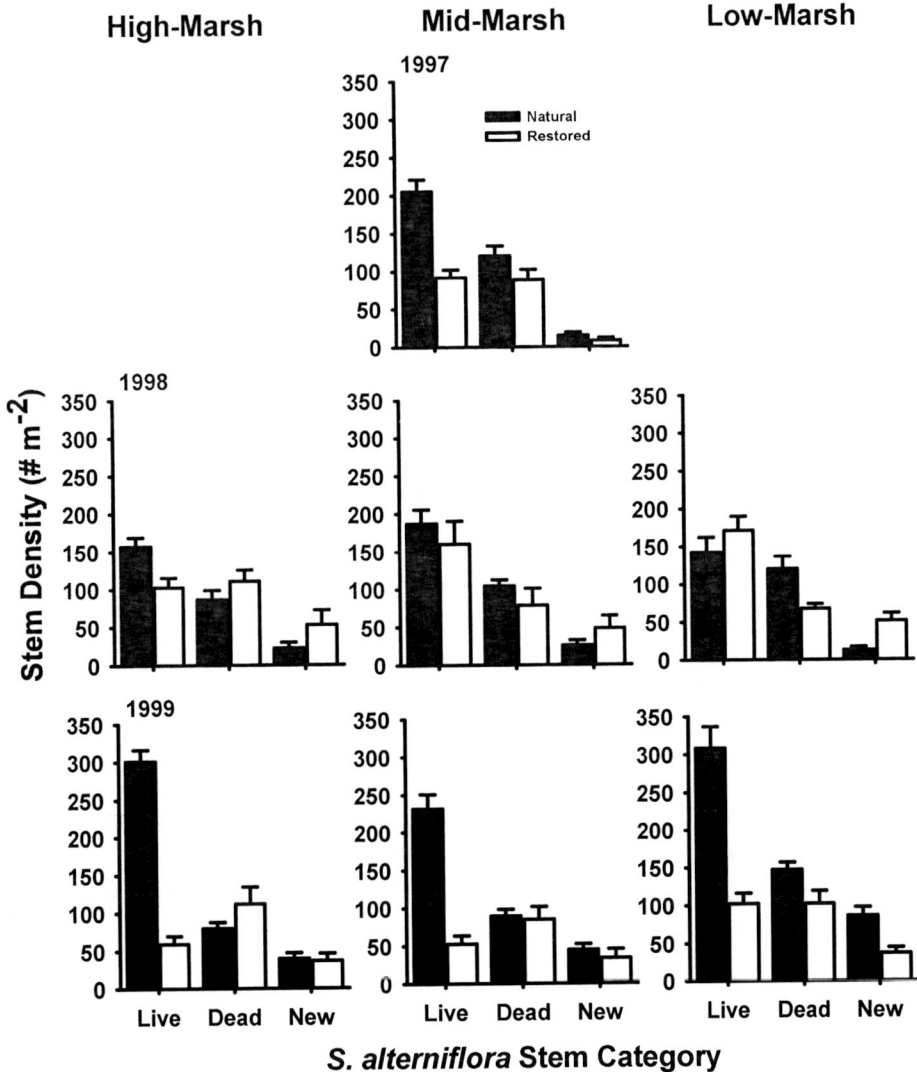

Figure 1. *Spartina alterniflora* live and standing-dead stems and new shoots (mean + 1 se) enumerated from 0.25 m² quadrats (n=10) haphazardly placed in a mid-marsh zone in May 1997 and low-, mid-, and high-marsh zones in May 1998 and 1999 within natural and restored Kamehameha marshes.

Mid-marsh stem heights were similar but nearest-neighbor distances differed between marshes. Natural, 82.4 ± 2.2 cm, compared to restored marsh live stems, 79.3 ± 2.4 cm, were slightly taller, not significantly different between marshes ($F_{1,\ 18} = 0.53$, $p > 0.05$), but significantly different among quadrats ($F_{18,\ 90} = 2.26$, $p < 0.007$). Nearest-neighbor distances were not affected by a marsh*stem type interaction ($F_{1,\ 96} = 0.02$, $p > 0.05$) or stem type ($F_{1,\ 96} = 3.61$, $p > 0.05$), but were significantly different between marshes ($F_{1,\ 96} = 21.55$, $p < 0.001$). Both live and dead stems were almost twice as far apart within the restored marsh (Fig. 2). Based on stem densities and nearest-neighbor distances natural marsh live ($\chi^2_{50} = 20.4$, $p > 0.05$) and dead stems ($\chi^2_{50} = 19.1$, $p > 0.05$) were randomly distributed while restored marsh live ($\chi^2_{50} = 83.8$, $p < 0.005$) and dead stems ($\chi^2_{50} = 76.0$, $p < 0.025$) were significantly clumped (Thompson 1956).

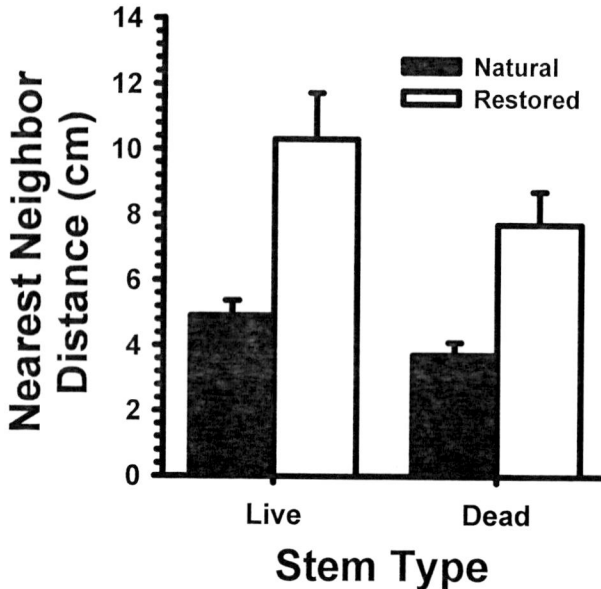

Figure 2. Natural and restored Kamehameha mid-marsh live and standing-dead stem mean (±1 se) nearest neighbor distances (n = 25) in May 1998.

Sediment parameters in a number of instances varied unpredictably by zone, date and marsh treatments. Pore water salinities were not significantly different between low- to high-marsh zones ($F_{2, 54}$ = 0.44, p > 0.05) unlike patterns typically observed within southeastern marshes (Pennings et al. 2005). Date*marsh ($F_{1, 54}$ = 1.60, p > 0.05) and marsh effects ($F_{1, 54}$ = 1.11, p > 0.05) were not significant, but salinities were significantly different between years ($F_{1, 54}$ = 41.39, p < 0.001). Salinities were reduced slightly in 1997 compared to 1998 (Fig. 3). The silt/clay sediment fraction was significantly different among zones ($F_{2, 54}$ = 23.52, p < 0.001) and between marshes ($F_{1, 54}$ = 112.52, p < 0.001), but date*marsh ($F_{1, 54}$ = 0.32, p > 0.05) and date effects ($F_{1, 54}$ < 0.001, p > 0.05) were not significant. Natural compared to restored marsh sediments contained greater percentages of silt/clay (Fig. 4). In the sandier restored marsh, the silt-clay content of low-marsh sediments was greater compared to sediments from other zones (Fig. 4). Sediment organic content was significantly different among zones ($F_{2, 52}$ = 6.92, p < 0.003) and between dates ($F_{1, 52}$ = 6.83, p < 0.015) and marshes ($F_{1, 52}$ = 138.72, p < 0.001), but the date*marsh interaction was not significant ($F_{1, 52}$ = 0.07, p > 0.05). In general organic content was greater in natural compared to restored marshes (Fig. 5). The collection of natural, high-marsh samples from within *J. roemerianus* stands in 1998 likely was a greater influence on AFDM date differences than actual year to year variation in sediment organic content. Similarly, the Chl *a* content of surface sediments was significantly different among zones ($F_{2, 53}$ = 18.55, p < 0.001) and between dates ($F_{1, 53}$ = 9.23, p < 0.005) and marshes ($F_{1, 53}$ = 15.82, p < 0.001), but the date*marsh interaction was not significant ($F_{1, 53}$ = 0.04, p > 0.05). Chl *a* amounts were twice as much in natural compared to restored sediments except in the high-marsh where amounts where almost equal (Fig. 6). The greater organic content of high-marsh natural sediments (Fig. 5) was not the result of differences in microalgal biomass between marshes.

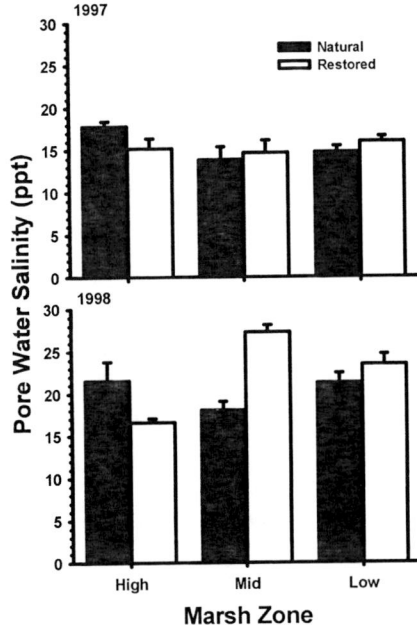

Figure 3. The mean (±1 se) pore water salinity measured from 0-2 cm depth sediment samples (n = 5) collected in July 1997 and June 1998 within low-, mid-, and high-marsh zones in natural and restored Kamehameha marshes.

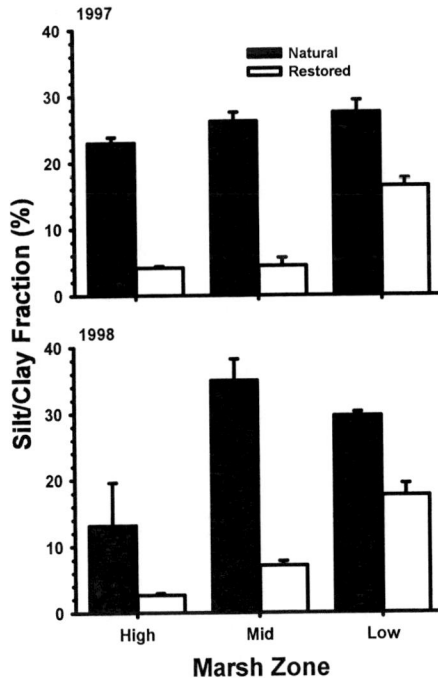

Figure 4. The mean (±1 se) percent silt-clay content (n = 5) of sediments collected from 0-2 depths within July 1997 and June 1998 low-, mid- and high-marsh zones in natural and restored Kamehameha marshes.

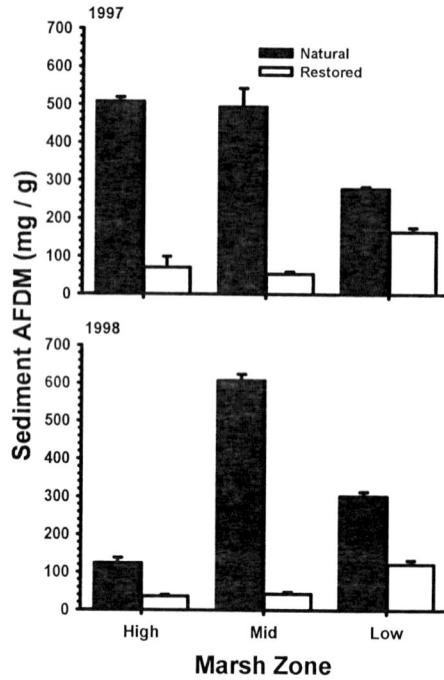

Figure 5. The mean (±1 se) sediment AFDM (n = 5) measured from 0-2 cm depths within July 1997 and June 1998 low-, mid- and high-marsh zones in natural and restored Kamehameha marshes.

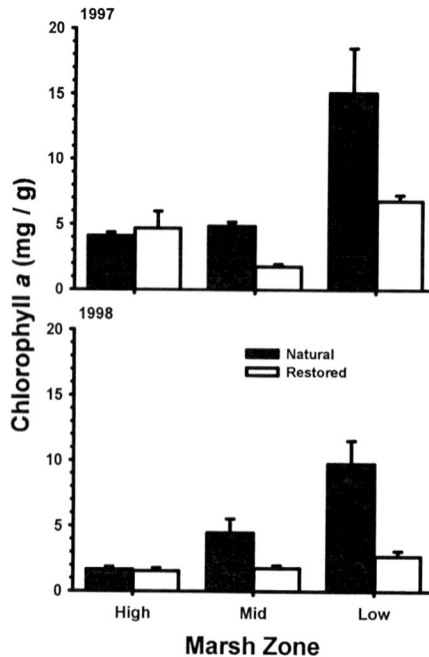

Figure 6. The mean (±1 se) chlorophyll *a* amounts (n = 5) in sediment samples collected from 0-2 cm depth in July 1997 and June 1998 low-, mid- and high-marsh zones in natural and restored Kamehameha marshes.

Growth and Feeding Experiments

Marsh Mussels

Densities of mussels collected within the field varied among zones and between dates and marsh (Fig. 7). Numbers were significantly different among zones ($F_{2, 114} = 11.49$, $p < 0.001$) and reflected a significant date*marsh interaction ($F_{1, 114} = 4.12$, $p < 0.05$). The natural marsh always had greater numbers of mussels except in 1999 when an influx of *G. demissa* resulted in similar densities between natural and restored mid-marsh locations (Fig. 7).

The haphazard allocation of mussels to enclosures in both 1997 and 1998 experiments resulted in few meaningful initial differences between zones and marshes. The average length of mussels deployed within enclosures in 1997 was 38.8 ± 0.5 mm (29.0 to 48.8 mm) and in 1998 was 58.3 ± 3.6 mm (32.1 to 78.2 mm). Shell lengths were significantly different between zones ($F_{1, 37} = 5.04$, $p < 0.04$) but not marshes in 1997 ($F_{1, 37} = 0.04$, $p > 0.05$). The actual mean difference between elevations was < 3.0 mm. In 1998 shell lengths were not significantly different by zones ($F_{1, 140} = 1.66$, $p > 0.05$), marsh ($F_{1, 28} = 0.86$, $p > 0.05$), or cages within marsh ($F_{28, 140} = 1.24$, $p > 0.05$).

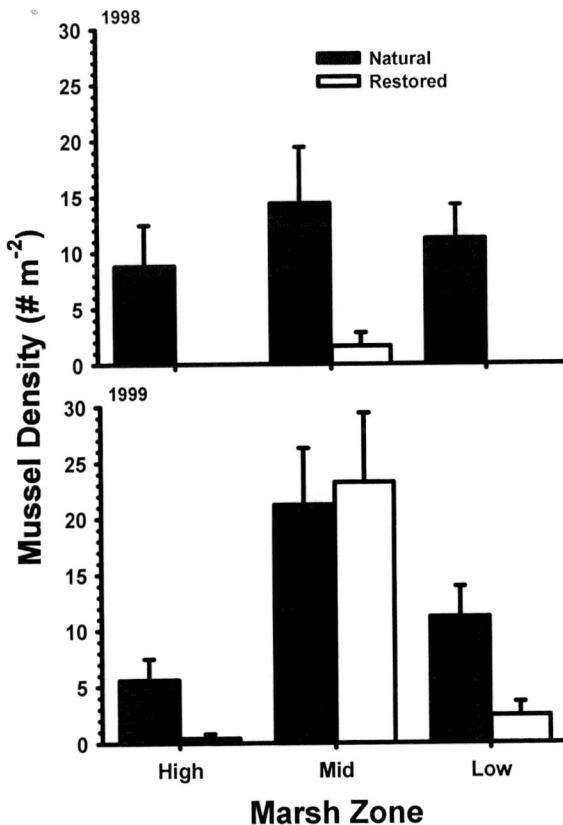

Figure 7. Mussel field densities (mean ± 1 se) in 1998 and 1999 from haphazardly placed 0.25 m^2 quadrats ($n = 10$) within low-, mid- and high-marsh zones in natural and restored Kamehameha marshes.

Initial shell and tissue dry mass for mussels used in growth experiments were calculated from power equation regression results for 1997 (n = 39) and 1998 initial field samples (n = 41). In 1997 shell length was a best-fit predictor of shell (r^2 = 0.86, $F_{1, 37}$ = 226.04, p < 0.001; dry mass = 0.056*(length)$^{2.779}$) and tissue dry mass (r^2 = 0.43, $F_{1, 37}$ = 27.74, p < 0.001; dry mass = 0.03*(length)$^{2.115}$). Shell length also was a significant predictor of shell (r^2 = 0.96, $F_{1, 39}$ = 927.5, p < 0.001; dry mass = 0.038*(length)$^{2.939}$) and tissue dry mass in 1998 (r^2 = 0.85, $F_{1, 39}$ = 215.3, p < 0.001; dry mass = 0.013*(length)$^{2.526}$).

In both 1997 and 1998 experiments, the growth in shell length and mass and tissue mass varied by marsh and elevation. Shell length and mass and tissue mass increased in a majority of the 65% of the mussels surviving in 1997, but tissue mass declined in a majority of the 64% of mussels surviving in 1998. The linear growth of shell was significantly different between zones and marshes in both 1997 and 1998 experiments (Table 1). Changes in shell length were greater in natural mid-marsh sites in 1997 but were greater in all restored marsh zones in 1998 (Fig. 8). Changes in shell mass were not appreciably different between zones and marshes in either 1997 or 1998 experiments (Table 1). Mussels generally added shell at a greater rate in 1998 and appreciably more shell was added in restored marsh mussels (Fig. 8). Changes in tissue dry mass were significantly different between zones with low-marsh mussel tissue generally increasing in both 1997 and 1998 experiments (Table 1). The interpretation of patterns in tissue growth are confounded by a general lack of tissue growth in 1998 and the reversal of order in 1997, especially since low-marsh values are based on a single observation (Fig. 8).

Table 1. Summary of ANOVA results for 1997 and 1998 enclosure experiments determining the block effect of elevation (low, mid) and the main effect marsh (natural, restored) and on the change in shell length and mass and tissue mass of mussels. Only restored low-marsh mussels had positive tissue growth in 1998 so analyses not included.

Experiment	Variable	Source of Variation	F	df	p
1997	Shell length	Elevation	13.76	1, 19	<0.002
		Marsh	14.56	1, 19	<0.002
	Shell mass	Elevation	0.81	1, 16	>0.05
		Marsh	0.84	1, 16	>0.05
	Tissue mass	Elevation	6.78	1, 15	<0.03
		Marsh	0.28	1, 15	>0.05
1998	Shell length	Elevation	33.37	1, 91	<0.001
		Marsh	63.72	1, 91	<0.001
	Shell mass	Elevation	3.33	1, 67	>0.05
		Marsh	3.90	1, 67	<0.053

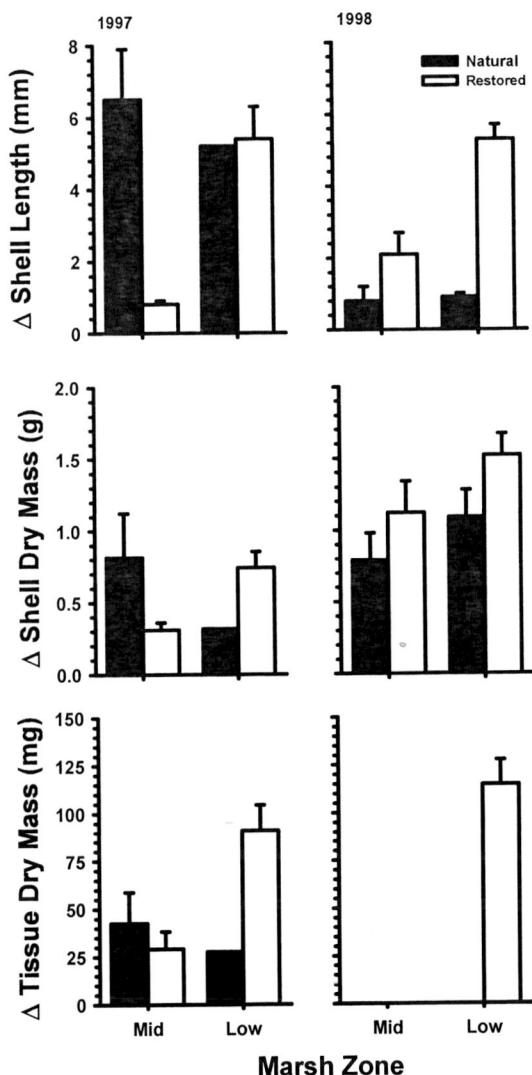

Figure 8. The mean (±1 se) yearly change in shell length and mass and tissue mass for mussels enclosed in mid- and low-marsh elevations within natural and restored Kamehameha marshes.

More suspended particulate matter, a potential source of food for mussels, settled out of the water column in the restored marsh but a greater percentage of the material that settled in the natural marsh was organic (Fig. 9). Settling particulates were not significantly different among zones ($F_{1, 35} = 2.94$, $p > 0.05$) but were affected by a significant date*marsh interaction ($F_{1, 35} = 6.82$, $p < 0.015$). Except for August mid-marsh samples, greater particulate amounts settled in the restored marsh (Fig. 9). The organic percentage of particulates that settled also were not significantly different between elevations ($F_{1, 35} = 0.07$, $p > 0.05$) and was not affected by a significant date*marsh interaction ($F_{1, 35} < 0.001$, $p > 0.05$). Particulate AFDM percents were significantly different between dates ($F_{1, 35} = 5.28$, $p < 0.03$) and marsh ($F_{1, 35} = 259.92$, $p < 0.001$). Although generally greater amounts of particulates settled in the restored marsh the overall AFDM of the particulates was significantly less during both sampling periods (Fig. 9).

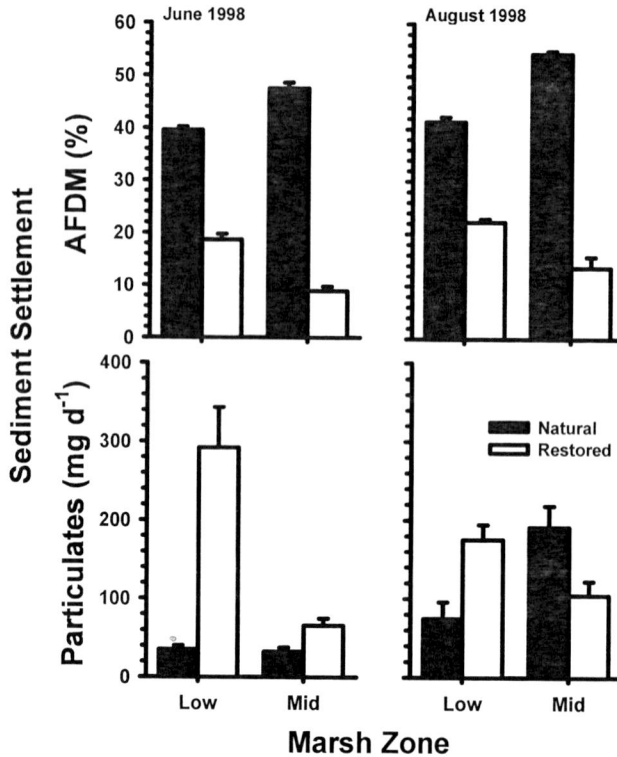

Figure 9. The mean (±1 se) daily amount of particulates and percent particulate AFDM (n = 5) for sediment settling out of the water column in July and August 1998 within low- and mid-marsh zones in natural and restored Kamehameha marshes.

Marsh Periwinkles

In both 1997 and 1998 experiments individuals were returned haphazardly to sites within marshes and the results were significant but few consequential initial differences existed between sites, dates and marshes. The average length of snails released in May 1997 was 16.3 ± 0.1 mm (9.8 to 20.4 mm) and May 1998 was 16.9 ± 3.6 mm (12.7 to 21.1 mm). Shell lengths were significantly different between sites ($F_{1, 793} = 4.43$, $p < 0.005$), dates ($F_{1, 37} = 73.06$, $p < 0.001$) and marshes ($F_{1, 793} = 8.30$, $p < 0.005$); there was no significant date*marsh interaction ($F_{1, 793} = 0.78$, $p > 0.05$). Actual mean differences between sites, dates or marshes were < 0.5 mm.

To calculate snail shell and tissue growth between May and August 1997, initial dry mass values were estimated from regression results for field collected individuals from the common St. Marys location (n = 69). Shell length was a best-fit predictor of shell ($r^2 = 0.79$, $F_{1, 68} = 260.4$, $p < 0.001$; dry mass = $1.113*(length)^{2.262}$) and tissue dry mass ($r^2 = 0.58$, $F_{1, 68} = 91.95$, $p < 0.001$; dry mass = $0.154*(length)^{2.183}$).

Snail recapture rates primarily varied between marshes. Recovery ranged between 35.5 to 37.0% for natural and 51.0 to 55.0% for restored marsh individuals in 1997 and 1998 experiments, respectively. Recapture rates were dependent on marsh ($\Delta\chi^2_1 = 14.38$, $p < 0.001$) and site within marsh ($\Delta\chi^2_1 = 26.96$, $p < 0.001$) but not year ($\Delta\chi^2_1 < 0.01$, $p > 0.5$). Visually locating snails within restored compared to natural marshes was easier (Walters pers. obs.).

The changes in snail wet mass, shell length and mass, and tissue mass varied by site, marsh and date. A majority of recaptured snails, >90%, increased in size and mass. The linear growth of shell was significantly different between sites, dates and marshes (Table 2). Shells grew more in 1997 than 1998 but consistently grew significantly more in the restored compared to the natural marsh (Fig. 10). The change in snail wet mass was significantly different between sites and marshes, but not dates (Table 2). Similar to length, snail wet mass was greater in the restored marsh (Fig. 10). Increased length and wet mass growth in restored marsh also translated into significantly greater increases in shell but not tissue dry mass in 1997 (Table 2). Shells increased by 68% in the restored compared to the natural marsh, but tissue mass was identical between marshes (Fig. 10).

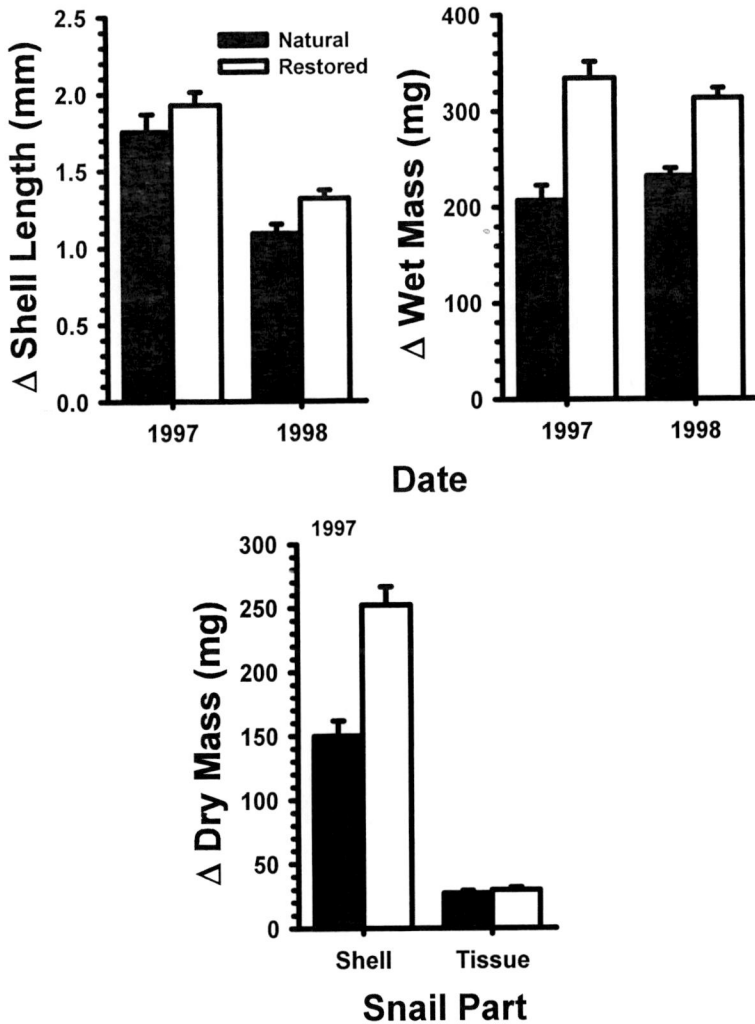

Figure 10. The mean (±1 se) May to August change in snail shell length and wet mass for 1997 and 1998 snail growth experiments and the change in shell and tissue dry mass for 1997 within natural and restored Kamehameha marshes.

Table 2. Summary of ANOVA results for 1997 and 1998 mark-recapture experiments determining the block effect of site and the main effects of date and marsh on the change in snail wet mass and shell length. Analyses of changes in shell and tissue mass also are included for 1997.

Experiment	Variable	Source of Variation	F	df	p
1997 & 1998	Shell length	Site	6.81	3, 341	<0.001
		Date	63.27	1, 341	<0.001
		Marsh	7.62	1, 341	0.007
		Date * Marsh	0.58	1, 341	>0.05
	Wet mass	Site	9.36	3, 341	<0.002
		Date	0.54	1, 341	>0.05
		Marsh	51.58	1, 341	<0.001
		Date * Marsh	1.31	1, 341	>0.05
1997	Shell mass	Site	2.73	3,151	<0.05
		Marsh	18.68	1, 151	<0.001
	Tissue mass	Site	8.03	3, 151	<0.001
		Marsh	0.06	1, 151	>0.05

Fiddler Crabs

The AFDM of sediments at the start of feeding and preference experiments was significantly different between marshes ($F_{1, 8} = 24.25$, $p < 0.002$). Natural marsh amounts (47.0 ± 2.4 mg g^{-1}) were >1.5X the AFDM in the restored marsh (28.2 ± 1.8 mg g^{-1}). Feeding and preference experiment crabs averaged 22.3 ± 0.7 mm in carapace width. In the feeding experiment significant container ($F_{16, 220} = 20.58$, $p < 0.001$) and marsh effects were identified ($F_{1, 16} = 36.99$, $p < 0.001$), but marsh*crab ($F_{1, 16} = 0.03$, $p > 0.05$) and crab effects ($F_{1, 16} < 0.01$, $p > 0.05$) were not significant. The AFDM increased from initial amounts in all treatments and almost doubled in both crab and no crab natural marsh sediments (Fig. 11). MANOVA identified significant container ($F_{8, 226} = 2.85$, $p < 0.006$) and crab treatment effects ($F_{2, 113} = 51.18$, $p < 0.001$) in the preference experiment. The presence of crabs appeared to increase AFDM in restored marsh sediments while decreasing amounts in natural marsh sediments (Fig. 11).

Survival Experiments

Mussel survival was dependent on marsh and zone. Individual losses in 1997 were influenced by marsh ($\Delta\chi^2_1 = 4.05$, $p < 0.05$) while losses in 1998 were dependent on both marsh ($\Delta\chi^2_1 = 12.75$, $p < 0.001$) and zone ($\Delta\chi^2_1 = 4.10$, $p < 0.05$). The effect of zone on survival, consistent with mortality in enclosure experiments not resulting from macropredators, tended to be greater in natural and mid-marsh locations (Fig. 12). However, survival always was greater in the restored marsh, between 70 to 90%, at all zones. Survival during the June to August 1998 tethering experiment, a measure of relative differences in macropredator mortality, was influenced by both marsh and zone effects.   Survival Experiments

Mussel survival was dependent on marsh and zone. Individual losses in 1997 were influenced by marsh ($\Delta\chi^2_1$ =4.05, p < 0.05) while losses in 1998 were dependent on both marsh ($\Delta\chi^2_1$ =12.75, p < 0.001) and zone ($\Delta\chi^2_1$ =4.10, p < 0.05). The effect of zone on survival, consistent with mortality in enclosure experiments not resulting from macropredators, tended to be greater in natural and mid-marsh locations (Fig. 12). However, survival always was greater in the restored marsh, between 70 to 90%, at all zones. Survival during the June to August 1998 tethering experiment, a measure of relative differences in macropredator mortality, was influenced by both marsh and zone effects.

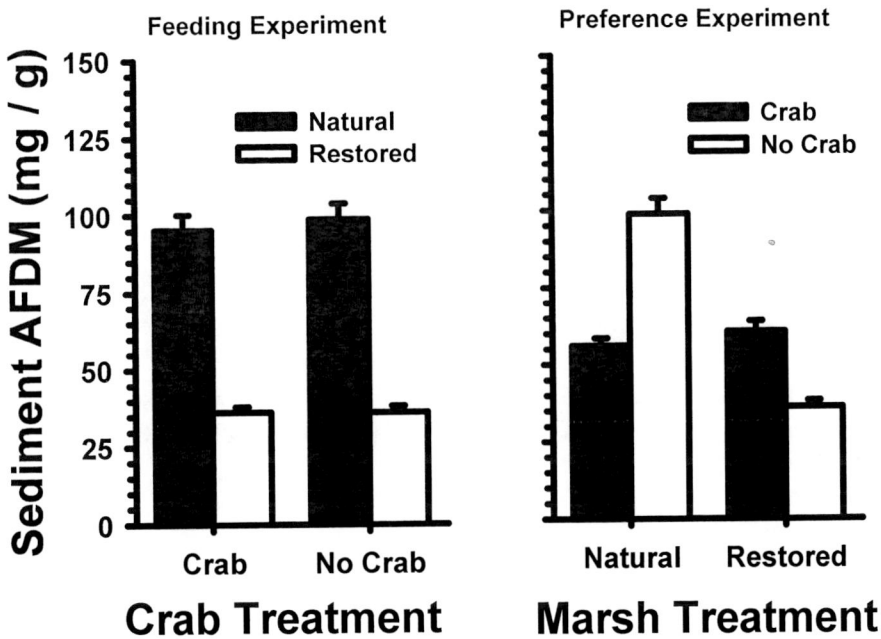

Figure 11. Crab effects on sediment AFDM (mean ±1 se) in feeding and preference experiments.

Only 1 mussel was lost during the first 5 d of observations. Survival after ca. 60 d was dependent on marsh ($\Delta\chi^2_1$ =4.09, p < 0.05) and zone ($\Delta\chi^2_1$ = 6.10, p < 0.015). Mid-marsh mussels had greater survival rates in either natural or restored marshes (Fig. 13), but survival was reduced in the restored marsh and was half that in the natural low-marsh (Fig 13). Mussel behavior appeared unaffected by tethering. All individuals were buried within the sediments similar to adjacent, untethered mussels and could be found within natural clusters of untethered mussels (Walters pers. obs.).

Snail survival during May - June and June – July tethering experiments was not influenced by marsh. No snails were lost during the first 5 d of each experiment. Fifty percent of the snails survived after ≈30 d in both natural and restored marshes in May – June. In the June – July experiment 30 d survival was dependent on zone ($\Delta\chi^2_1$ = 8.42, p < 0.004) but not marsh ($\Delta\chi^2_1$ = 2.53, p > 0.5).

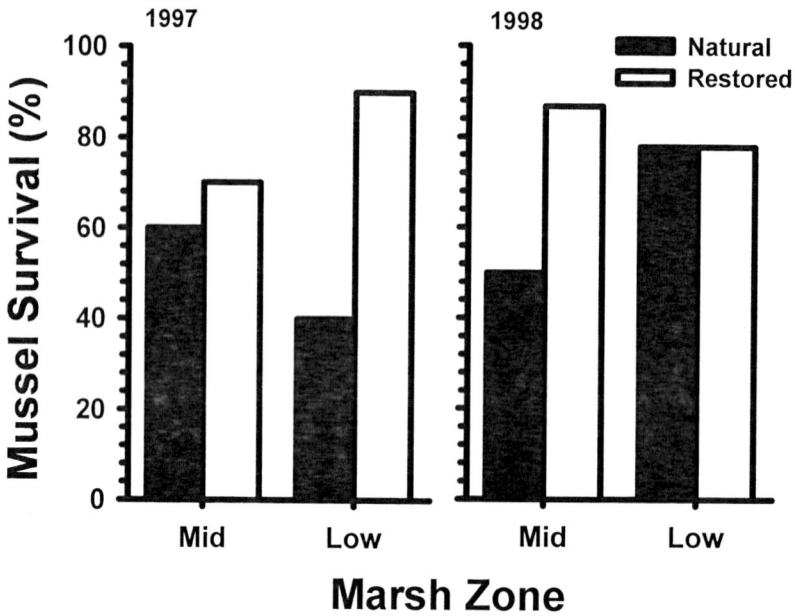

Figure 12. The percent survival of mussels deployed in 1997 and 1998 enclosure experiments.

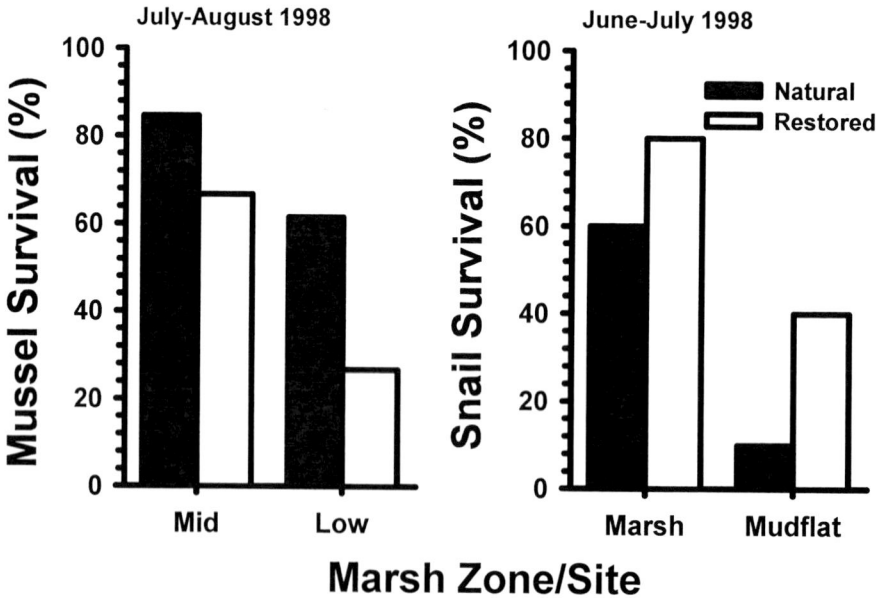

Figure 13. The percent survival of mussels or snails deployed in 1998 tethering experiments.

Mortality for snails tethered on the mud flat was greater than for snails within the marshes (Fig. 13). Tethering snails did not appear to prevent typical movements up and down stems and except for a few instances when individuals became entangled and suspended off the sediment appeared to have a minimal effect on behavior (Walters pers. obs.).

DISCUSSION

Significant structural differences existed between the Kamehameha marshes during the first 3-4 yrs since the creation of the restored marsh. An early parity in 1998 stem densities was replaced in the last year of the study by a >3 fold increase in low-marsh numbers within natural compared to restored marshes. Plant biomass typically is one of the first attributes to recover in marshes but can take anywhere from 14 mo. up to 10 yrs. to equal natural marsh amounts (Craft et al. 1999, Laegdsgaard 2002). The similar Kamehameha stem densities in 1998 was rapid, but the 1999 increases in natural marsh densities coupled with a decline in restored marsh numbers, especially given the greater shoot densities the year before, is not readily explained. The spatial arrangement of plants in restored mid-marsh sites also resulted in greater distances among stems, but unexpectedly canopy architecture was similar between marshes in only the second growing season. Stem distances likely still reflect the grid-pattern of initial *S. alterniflora* plug planting, but the similar canopy heights are distinct from past studies where equivalency can take many years (Zedler 1993).

Physical differences in soil characteristics are consistent with previous studies of restored marshes and reflect the often decades necessary to accumulate organic matter (Craft et al. 1999). The 2-5 fold differences in silt/clay fraction and sediment AFDM occur in both years and all marsh zones. Except in low-marsh sites, the biomass of microphytobenthos is not responsible for observed marsh and zone differences in organic matter. Chlorophyll *a* levels only are greater for natural low-marsh sites and likely contribute to the natural vs. restored marsh differences in AFDM. The early similarity in Chl *a* amounts does not conform with results from North Carolina marshes where restored marsh age significantly affected algal biomass (Zheng et al. 2002). Total particulate settlement tended to be greater within the restored marsh and, although measured for only a relatively short period of time, also does not explain the greater AFDM in natural marsh sediments. However, the AFDM of settling particulates was consistently lower and could contribute to a slower accumulation of organics in the restored marsh. Pore water salinities are unremarkably similar between marshes but did not exhibit a typical increase from low- to high-marsh zones (Pennings et al. 2005). The normal evaporation experienced in high-marsh zones resulting in increased salinity may not occur in the Kamehameha marshes. Both marshes are located in close proximity, <90 m, to very large, unused dredge spoil catchments that were filled with water. Observations suggest that water from the catchments may leach into the marshes preventing the high-marsh areas from experiencing normal drying.

An initial absence of large macrofaunal residents in the restored marsh is consistent with previous marsh restoration studies where often 10+ yrs. are necessary before faunal parity is reached (Posey et al. 1997, Craft et al. 1999, Talley & Levin 1999, Thelen & Thiet 2008). The rapidity with which mussel densities in the restored marsh increased, especially in the mid-marsh, suggests that at least mussel recovery may be occurring more rapidly than anticipated. A total absence of *L. irrorata,* other than ones transplanted from the St. Marys site, in both natural and restored marshes was unique. The snail absence suggests the upper reaches of the North River where both restored and natural marshes are located may experience less than typical larval settlement of certain marsh resident species and reinforces the observation that modes of colonization can have a major effect on marsh faunal recovery (Craft & Sacco 2003, Moseman et al. 2004).

Experiments assessing functional similarity between Kamehameha natural and restored marshes generally contradict the overall conclusion that recovery has not occurred based on existing structural differences (e.g., stem densities, organic matter accumulation, faunal presence). Mussel non-predatory survival and, except for 1997 mid-marsh mussels, growth were greater in restored compared to natural marshes. Restored marsh mussel shell length and dry mass and tissue dry mass increased by >100% in a number of instances and were the only mussels adding tissue mass in 1998. Inundation frequency and water column food availability affect the growth of marsh mussels (Franz 1997), and differences in relative elevation between natural and restored mid-marsh sites could explain the greater 1997 growth of natural marsh mussels. However, detailed field maps indicate enclosures were placed in close proximity each year, though exact elevations were not determined. The reversal of growth patterns between years, restored greater than natural in 1998, also suggests some factor other than elevation is responsible for year to year differences. Snails that rely directly on extant and not tidally transported marsh resources (Bärlocher & Newell 1994, Silliman & Newell 2003) grew better in the restored marsh during both summers. Increases in shell length and mass primarily accounted for the 1997 difference in summertime growth, but molluscs frequently exhibit temporal disconnects between the growth of shell and tissue (e.g., Borrero & Hilbish 1988). Fiddler crabs demonstrated no difference in feeding on either natural or restored marsh sediments but did seem to prefer natural marsh sediments when offered a choice. Sediment characteristics (Robertson & Newell 1982, Jaramillo & Lunecke 1988) along with temperature and salinity regimes (Frusher et al. 1994) can affect intertidal *Uca* spp. distributions, but fiddler growth rates appear to vary little among different habitats (Koch et al. 2005). The feeding preference exhibited by crabs in this study also may be more apparent than real. In the crab addition treatment natural and restored marsh sediment AFDM were equivalent, but the AFDM in restored marsh sediments consistently are significantly lower in all other samples. One possible explanation is that crab behavior may redistribute organic matter across the two sediment types within the containers confounding the ability to detect a feeding preference.

Although similar to or better than natural marshes for feeding, growth and non-predatory mortality, the restored marsh was an inconsistent refuge from predators. Mussel losses to predators were ca. 25-67% greater in the restored marsh, but snail losses were just the opposite and were greater in the natural marsh. Vegetated habitats typically provide a refuge against numerous macro-predators (Leber 1985, Halpin 2000), but the mechanism can be complex involving intraguild as well as predator-structure interactions (Grabowski & Powers 2004, Griffen & Byers 2006). In the Kamehameha marsh system the clumped distribution of stems within the restored marsh likely resulted in a less effective inhibition for predators feeding on mussels. However, the ability of snails also to crawl up *S. alterniflora* stems (Hovel et al. 2001) provided an additional, effective predator avoidance behavior.

The ability of salt marsh restoration efforts to deliver marshes similar in ecological function remains an open question. Depending on the goal and metrics applied most marsh restoration projects could be considered qualified successes. The equivalency of plant biomass (Broome et al. 1986, Roman et al. 2002), genetic diversity (Travis et al. 2002), fish utilization (Roman et al. 2002, Able et al. 2004), and pore water salinities (this study) to name just a few can be established rapidly in restored marshes. Other metrics including sediment organic content (Craft et al. 1999), nutrient cycling (Thompson et al. 1995, Craft et al. 1999), faunal density and richness (Sacco et al. 1994, Levin et al. 1996), and plant canopy architecture (Zedler 1993) tend to require significant time if not decades before reaching parity with natural marshes. Although each of the above metrics contributes either individually or in concert to ecological function, comparisons between marshes do not address directly the question of functional similarity. In this study experiments designed to test for differences in the primary marsh functions of providing food for consumption and/or a refuge from predation indicate that in a radical marsh restoration project, where pine upland was converted to intertidal elevations, functional equivalency between marshes can occur in <4 yrs. Experimental results were not unequivocal, the restored marsh was not as effective a refuge for mussels, but did provide insights into the potential reasons for any lack of functional equivalency (e.g., stem nearest-neighbor distances).

Continued calls for more rigorously designed projects (Chapman & Underwood 2000, Grayson et al. 1999) and more effective integration of the fields of ecological restoration and restoration ecology (Weinstein 2007) are resulting in identification of improved analytic approaches (Holl et al. 2003), the use of case studies to guide new restoration efforts (Hopfensperger et al. 2006), long-term and regional studies (Craft et al. 1999, Warren et al. 2002, Teal & Peterson 2005, Konisky et al. 2006), and the application of experimental approaches (Lindig-Cisneros & Zedler 2002, Zedler & Callaway 2003) to gauge the overall success of salt marsh restoration efforts. Targeted experiments addressing directly known marsh functions can provide invaluable insights into the success or failure of restoration projects. Experiments, while not inexpensive, can provide a cost effective alternative to time consuming and expensive monitoring efforts frequently mandated by regulatory agencies.

ACKNOWLEDGMENTS

This study would not have been possible without the unstinting assistance of P. Schoenfeld, R. Wilkenson and J. Garner (Kings Bay) for facilitating access to field sites and base accommodations. Many thanks to S. Pennings for design inspirations; K. Barrett, S. Brandon, N. Gann, M. Jones, M. Neftzger, M. Nunley, and D. Suntheimer for lab and field assistance; D. Curry and M. Norman for support; the Cluster in Applied Ecology and Conservation at CCU and anonymous reviewers for critical evaluation and constructive comments on earlier drafts. This research was supported in part by the TN Board of Regents Technology Access Program, Middle TN State University's Graduate Research Council, a College of Basic and Applied Science Research Award and the Department of Biology Scholarship Program and the US Navy, Kings Bay Submarine Base, Facilities and Environmental Division.

REFERENCES

Able, K. W., Nemerson, D. M. & Grothues, T. M. (2004). Evaluating salt marsh restoration in Delaware Bay: Analysis of fish response at former salt hay farms. *Estuaries*, 27, 58-69.

Allen, E. A., Fell, P. E., Peck, M. A., Gieg, J. A., Guthke, C. R., & Newkirk, M.D. (1994). Gut contents of common mummichogs, *Fundulus heteroclitus* L., in a restored impounded marsh and a natural reference marsh. *Estuaries,* 17, 462-471.

Aronson, R.B. & Heck, K.L. (1995). Tethering experiments and hypothesis testing in ecology. *Mar. Ecol. Prog. Ser.*, 121, 307-309.

Aronson, R.B., Heck, K.L. & Valentine, J.F. (2001). Measuring predation with tethering experiments. *Mar. Ecol. Prog. Ser.*, 214, 311-312.

Bärlocher, F. & Newell, S. Y. (1994). Growth of the salt marsh periwinkle *Littoraria irrorata* on fungal and cordgrass diets. *Mar. Biol.*, 118, 109-114.

Baustian, J. J. & Turner, R. E. (2006). Restoration success of backfilling canals in coastal Louisiana marshes. *Restor. Ecol.*, 14, 636-644.

Bertness, M. D. & Grosholz, E. (1985). Population dynamics of the ribbed mussel, *Geukensia demissa*: the costs and benefits of an aggregated distribution. *Oecologia*, 67, 192-204.

Bortolus, A., Schwindt, E. & Iribarne, O. (2002). Positive plant-animal interactions in the high marsh of an Argentinean coastal lagoon. *Ecology*, 83, 733-742.

Borrero, F. J. & Hilbish, T. J. (1988). Temporal variation in shell and soft tissue growth of the mussel *Geukensia demissa*. *Mar. Ecol. Prog. Ser.*, 42, 9-15.

Brady, V., Cardinale, B., Gathman, J. & Burton, T. (2002). Does facilitation of faunal recruitment benefit ecosystem restoration? An experimental study of invertebrate assemblages in wetland mesocosms. *Restor. Ecol.*, 10, 617-626.

Broome, S. W. (1990). Creation and restoration of tidal wetlands of the Southeastern United States. In: J. A. Kunsler & M. E. Kentula (Eds.), *Wetland creation and restoration*. Washington, DC: Island Press.

Broome, S. W., Seneca, E. D. & Woodhouse, W. W. (1982). Establishing brackish marsh on graded upland sites in North Carolina. *Wetlands*, 2, 152-178.

Broome, S. W., Seneca, E. D. & Woodhouse, W. W. (1986). Long-term growth and development of transplants of salt-marsh grass *Spartina alterniflora*. *Estuaries*, 9, 63-74.

Burnham, K. P. & Anderson, D. R. (1998). Model selection and inference. *N.Y., N.Y.*: Springer.

Carpenter, S. R., Kitchell, J. R., Cottingham, K. L., Schindler, D. E., Christensen, D. L., Post, D. M. & Voichick, N. (1996). Chlorophyll variability, nutrient input, and grazing: evidence from whole-lake experiments. *Ecology*, 77, 725-735.

Chapman, M. G. (1999) Improving sampling designs for measuring restoration in aquatic habitats. *J. Aquatic Ecosystem Stress & Recovery*, 6, 235-251.

Chapman, M. G. & Underwood, A. J. (2000). The need for a practical scientific protocol to measure successful restoration. *Wetlands (Australia)*, 19, 28-49.

Clark, K. L., Ruiz, G. M. & Hines, A. H. (2003). Diel variation in predator abundance, predation risk and prey distribution in shallow-water estuarine habitats. *J. Exp. Mar. Biol. Ecol.*, 287, 37-55.

Clark, V. (2004). SAS/STAT 9.1 Users' Guide. Carey, *NC: SAS Publishing*.

Costanza, R., d'Arge, R., de Groot, R., Farberk, S., Grasso, M., Hannon, B., Limburg, K., Naeem, S., O'Neill, R. V., Paruelo, J., Raskin, R.G., Suttonkk, P., & van den Belt, M., (1997). The value of the world's ecosystem services and natural capital. *Nature*, 387, 253-260.

Craft, C. & Sacco, J. N. (2003). Long-term succession of benthic infauna communities on constructed *Spartina alterniflora* marshes. *Mar. Ecol. Prog. Ser.*, 257, 45-58.

Craft, C. B., Broome, S. W. & Seneca, E. D. (1988). Nitrogen, phosphorus and organic carbon pools in natural and transplanted marsh soils. *Estuaries*, 11, 272-280.

Craft, C. B., Seneca, E. D. & Broome, S. W. (1991). Porewater chemistry of natural and created marsh soils. *J. Exp. Mar. Biol. & Ecol.*, 152, 187-200.

Craft, C.,Reader, J., Sacco, J. N. & Broome, S. W. (1999). Twenty-five years of ecosystem development of constructed *Spartina alterniflora* (Loisel) marshes. *Ecol. Appl.*, 9, 1405-1419.

Currin, C. A., Joye, S. B. & Paerl, H. W. (1996). Diel rates of N_2-fixation and denitrification in a transplanted *Spartina alterniflora* marsh: implications for N-flux dynamics. *Est., Coast. & Shelf Sci.*, 42, 597-616.

Dame, R.F. (1989). The importance of *Spartina alterniflora* to Atlantic Coast estuaries. *Reviews in Aquatic Sciences*, 1, 639-660.

Dobson, A. P., Bradshaw, A. D. & Baker, A. J. M. (1997). Hopes for the Future: restoration ecology and conservation biology. *Science,* 277, 515-522.

Dixon, P. & Ellison, A. M. (1996). Introduction: Ecological applications of Bayesian influence. *Ecol. Appl.* 6, 1034-1035.

Eggleston, D. B., Bell, G. W. & Amavisca, A. D. (2005). Interactive effects of episodic hypoxia and cannibalism on juvenile blue crab mortality. *J. Exp. Mar. Biol. Ecol.*, 325, 18-26.

Franz, D. R. (1997). Resource allocation in the intertidal salt-marsh mussel *Geukensia demissa* in relation to shore level. *Estuaries*, 20, 134-248.

Frusher, S. D., Giddins, R. L. & Smith, T. J. (1994). Distribution and abundance of grapsid crabs in a mangrove estuary: effects of sediment characteristics, salinity tolances and osmoregulatory ability. *Estuaries*, 17, 647-654.

Gallagher, J. L., Somers, G. F., Grant, D. M., & Seliskar, D. M. (1988). Persistent differences in tow forms of *Spartina alterniflora*: a common garden experiment. *Ecology,* 69, 1005-1008.

Grabowski, J. H. & Powers, S. P. (2004). Habitat complexity mitigates trophic transfer on oyster reefs. *Mar. Ecol. Prog. Ser.*, 277, 291-295.

Grayson, J. E., Chapman, M. G. & Underwood, A. J. (1999). The assessment of restoration of habitat in urban wetlands. *Landscape & Urban Plan.*, 43, 227-236.

Griffen & Byers (2006). Partitioning mechanisms of predator interference in different habitats. *Oecologia*, 146, 609-614.

Halpin, P. M. (2000). Habitat use by an intertidal salt-marsh fish: trade-offs between predation and growth. *Mar. Ecol. Prog. Ser.*, 198, 203-214.

Heck, K. & Wilson, K. (1987). Predation rates on decapod crustaceans in latitudinally separated seagrass communities: a study of spatial and temporal variation using tethering techniques. *J. Exp. Mar. Biol. Ecol.,* 107, 87-100.

Higgs, E. S. (2005). The two-culture problem: Ecological restoration and the integration of knowledge. *Restor. Ecol.*, 13, 159-164.

Hilborn, R. & Mangel, M. S. (1997). The ecological detective. Princeton, *N. J.:* Princeton University Press.

Hilderbrand, R. H., Watts, A. C. & Randle, A. M. (2005). The myths of restoration ecology. *Ecology & Society*, 10, 19-29.

Holl, K. D., Crone, E. E. & Schultz, C. B. (2003). Landscape restoration: moving from generalities to methodologies. *BioScience,* 53, 491-502.

Hopfensperger, K. N., Engelhardt, K. A. M. & Seagle, S. W. (2006). The use of case studies in establishing feasibility for wetland restoration. *Restor. Ecol.*, 14, 578-586.

Hovel, K. A., Bartholomew, A. & Lipcius, R.N. (2001). Rapidly entrainable tidal vertical migrations in the salt marsh snail *Littoraria irrorata*. *Estuaries,* 24, 808-816.

Hurlbert, S.H. (1984). Pseudoreplication and the design of ecological field experiments. *Ecol. Mono.*, 54, 187-211.

Jaramillo, E. & Lunecke, K. (1988). The role of sediments in the distribution of *Uca pugilator* and *Uca pugnax* in a salt marsh of Cape Cod. *Meeresforsch,* 32, 46-52.

Keer, G. H. & Zedler, J. B. (2002). Salt marsh canopy architecture differs with the number and composition of species. *Ecol. Appl.*, 12, 456-473.

Kemp, P. F., Newell, S. Y. & Krambeck, C. (1990). Effects of filter-feeding by the ribbed mussel Geukensia demissa on the water-column microbiota of a Spartina alterniflora saltmarsh. *Mar. Ecol. Prog. Ser.*, 59, 119-131.

Kneib, R. T. & Scheele, C. E. H. (2000). Does tethering of mobile prey measure relative predation potential? An empirical test using mummichogs and grass shrimp. *Mar. Ecol. Prog. Ser.*, 198, 181-190.

Kock, V., Wolff, M. & Diele, K. (2005). Comparative population dynamics of four fiddler crabs (Ocypodidae, genus *Uca*) from a North Brazilian mangrove ecosystem. *Mar. Ecol. Prog. Ser.*, 291, 177-188.

Konisky, R. A., Burdick, D. M., Dionne, M. & Neckles, H. A. (2006). A regional assessment of salt marsh restoration and monitoring in the Gulf of Maine. *Rest. Ecol.*, 14, 516-525.

Kreeger, D. A. & Newell, R. I. E. (2001). Seasonal utilization of different seston carbon sources by the ribbed mussel, *Geukensia demissa* (Dillwyn) in a mid-Atlantic salt marsh. *J. Exp. Mar. Biol. Ecol.*, 260, 71-91.

Laegdsgaard, P. (2002). Recovery of small denuded patches of the dominant NSW coastal saltmarsh species (*Sporobolus virginicus* and *Sarcocornia quinqueflora*) and implication for restoration using donor sites. *Ecol. Manag. Restor.*, 3, 202-206.

Laegdsgaard, P. (2006). Ecology, disturbance and restoration of coastal saltmarsh in Australia: a review. *Wetlands Ecol. & Manag.*, 14, 379-399.

Langis, R., Zalejko, M. & Zedler, J. B. (1991). Nitrogen assessments in a constructed and a natural saltmarsh of San Diego Bay. *Ecol. Appl.*, 1, 40-51.

Leber, K. (1985). The influence of predatory decapods, refuge, and microhabitat selection on seagrass communities. *Ecology*, 66, 1951–1964.

Lent, C. M. (1969). Adaptations of the ribbed mussel, *Modiolus demissus* (Dillwyn) to the intertidal habitat. *Am. Zool.*, 9, 283-292.

Levin, L. A., Talley, D. & Thayer, G. (1996). Succession of macrobenthos in a created salt marsh. *Mar. Ecol. Prog. Ser.*, 141, 67-82.

Lindig-Cisneros, R. & Zedler, J. B. (2002). Halophyte recruitment in a salt marsh restoration site. *Estuaries,* 25, 1174-1183.

Lockwood, J. R. (1998). On the statistical analysis of multiple-choice feeding preference experiments. *Oecologia*, 116, 475-481.

Luckenbach, M.W., Coen, L.D., Ross Jr., P.G. & Stephen, J. A. (2005). Oyster reef habitat restoration: relationships between oyster abundance and community development based on two studies in Virginia and South Carolina. *J. Coastal Res.*, SI40, 64-78.

Luken, J. O. & Walters, K., (2009). Management of plant invaders within a marsh: an organizing principle for ecological restoration? In Inderjit (Ed.), *Management of invasive weeds* (pp.61-76). N.Y., N.Y.: Springer.

Manly, B. F. J. (1993). Comments on design and analysis of multiple-choice feeding-preference experiments. *Oecologia*, 93, 149-152.

Matthews, G. A. & Minnello, T. J., (1994). Techniques and success in restoration, creation and enhancement of *Spartina alterniflora* marshes in the United States. Vol. 1 - Executive summary and annotated bibliography. *In NOAA Coastal Ocean Program Decision Analysis Series (N. 2)*. Silver Springs, MD: NOAA Coastal Ocean Office.

Minello, T. J. & Zimmerman, R. J. (1992). Utilization of natural and transplanted Texas saltmarshes by fish and decapod crustaceans. *Mar. Ecol. Prog. Ser.*, 90, 273-285.

Minello, T. J. & Webb, J. W. (1997). Use of natural and created *Spartina alterniflora* salt marshes by fishery species and other aquatic fauna in Galveston Bay, Texas, USA. *Mar. Ecol. Prog. Ser.*, 151, 165-179.

Moseman, S. M., Levin, L. A., Currin, C. & Forder, C. (2004). Colonization, succession, and nutrition of macrobenthic assemblages in a restored wetland at Tijuana Estuary, California. *Est. Coast. Shelf Sci.*, 60, 755-770.

Moy, L. D. & Levin, L. A. (1991). Are *Spartina* marshes a replaceable resource? A functional approach to evaluation of marsh creation efforts. *Estuaries, 14*, 1-16.

Newell, S. Y. & Barlocher, F. (1993). Removal of fungal and total organic matter from decaying cordgrass leaves by shredder snails. *J. Exp. Mar. Biol. Ecol.*, 171, 39-49.

Nixon, S. W. (1980). Between coastal marshes and coastal waters - a review of twenty years of speculation and research on the role of salt marshes in estuarine productivity and water chemistry. In: P. Hamilton & K.B. MacDonald (Eds.), *Estuarine and wetlands processes with emphasis on modeling*, N.Y., N.Y.: Plenum Press.

Norušis, M. J. (2008) SPSS 16.0 Statistical Procedures Companion, *Upper Saddle River*, NJ: Prentice Hall.

Orth, R. J., Kendrick, G. A. & Marion, S. R. (2007). *Posidonia australis* seed predation in seagrass habitats of Two Peoples Bay, Western Australia. Aquatic Botany, 86, 83-85.

Parsons, T. R., Maita, Y. & Lalli, C.M. (1984). *A manual of chemical and biological methods for seawater analysis*. Oxford, U.K.: Pergamon Press.

Pennings, S. C., Grant, M., & Bertness, M.D. (2005). Plant zonation in low-latitude salt marshes: disentangling the roles of flooding, salinity and competition. *J. Ecology, 93*, 159-167.

Peterson, C. H. & Renaud, P. E. (1989). Analysis of feeding preference experiments. *Oecologia*, 80, 82-86.

Peterson, C. H. & Black, R. (1994). An experimentalist's challenge: when artifacts of intervention interact with treatments. *Mar. Ecol. Prog. Ser.*, 111, 289-297.

Peterson, C. H. & Lipcius, R. N., (2003). Conceptual progress towards predicting quantitative ecosystem benefits of ecological restorations. *Mar. Ecol. Prog. Ser*, 264, 297-307.

Piehler, M. F., Currin, C. A., Casanova, R. & Paerl, H. W. (1998). Development and N_2-fixing activity of the benthic microalgal community in transplanted *Spartina alterniflora* marshes in North Carolina. *Restor. Ecol.*, 6, 290-296.

Posey, M. H., Alphin, T. D. & Powell, C. M. (1997). Plant and infaunal communities associated with a created marsh. *Estuaries*, 20, 42-47.

Roa, R. (1992). Design and analysis of multiple-choice feeding-preference experiments. *Oecologia,* 89, 509-515.

Robertson, J. R. & Newell, S. Y. (1982). Experimental studies of particle ingestion by the sand fiddler crab *Uca pugilator. J. Exp. Mar. Biol. Ecol.*, 59, 1-21.

Roman, C. T., Raposa, K. B., Adamowicz, S. C., James-Pirri, M.-J. & Catena, J. G., (2002). Quantifying vegetation and nekton response to tidal restoration of a New England salt marsh. *Rest. Ecol.*, 10, 450-460.

Rozas, L. P. & Odum, W. E. (1988). Occupation of submerged aquatic vegetation by fishes: testing the roles of food and refuge. *Oecologia*, 77, 101-106.

Sacco, J. N., Seneca, E. D. & Wentworth, T. R. (1994). Infaunal community development of artificially established saltmarshes in North Carolina. *Estuaries*, 17, 489-500.

Seneca, E. D., Broome, S. W., & Woodhouse, W. W. (1985). The influence of duration-of-inundation on the development of a man-initiated *Spartina alterniflora* Loisel marsh in North Carolina. *J. Exp. Mar. Biol. & Ecol.*, 94, 259-268.

Short, F. T., Burdick, D. M., Short, C. A., Davis, R. C. & Morgan, P. A. (2000). Developing success criteria for restored eelgrass, salt marsh and mud flat habitats. *Ecol. Eng.*, 15, 239-252.

Silliman, B. R. & Newell, S. Y. (2003). Fungal farming in a snail. *Proc. Nat. Acad. Sci.*, 100, 15643-15648.

Simenstad, C. A. & Thom, R. M. (1996). Functional equivalency trajectories of the restored Gog-Le-Hi-Te estuarine wetland. *Ecol. Appl.*, 6, 38-56.

Talley, T. S. & Levin, L. A. (1999). Macrofaunal succession and community structure in *Salicornia* marshes of southern California. *Est. Coast. Shelf Sci.*, 49, 713-731.

Teal, J. M. & Peterson, S. B, (2005). Introduction to the Delaware Bay salt marsh restoration. *Ecol. Eng.*, 25, 199-203.

Thelen, B. A. & Thiet, R. K. (2008). Molluscan community recovery following partial tidal restoration of a New England estuary, *U.S.A. Restor. Ecol.*,

Thompson, H. R., (1956). Distribution of distance to n[th] neighbor in a population of randomly distributed individuals. *Ecology*, 37, 391-394.

Thompson, S. P., Paerl, H. W. & Go, M. C. (1995). Seasonal patterns of nitrification and denitrification in a natural and a restored salt marsh. *Estuaries*, 18, 399-408.

Travis, S. E., Proffitt, C. E., Lowenfeld, R. C. & Mitchell, T. W. (2002). A comparative assessment of genetic diversity among differently-aged populations of *Spartina alterniflora* on restored versus natural wetlands. *Restor. Ecol.*, 10, 37-42.

Underwood, A. J. (1994). On Beyond BACI: sampling design that might reliably detect environmental disturbances. *Ecol. Appl.*, 4, 3-15.

Vaughn, C. C. & Fisher, F. M. (1988). Vertical migration as a refuge from predation in intertidal marsh snails: a field test. *J. Exp. Mar. Biol. Ecol.*, 123, 163-176.

Walters, K. & Coen, L. D. (2006). A comparison of statistical approaches to analyzing community convergence between natural and constructed oyster reefs. *J. Exp. Mar. Biol. Ecol.*, 330, 81-95.

Warren, R. S., Fell, P. E., Rozsa, R., Brawley, A. H., Orsted, A. C., Olson, E. T., Swamy, V., Niering, W. A. (2002). Salt marsh restoration in Connecticut: 20 years of science and management. *Rest. Ecol.*, 10, 497-513.

Weinstein, M. P., Balletto, J. H., Teal, J. M. & Ludwig, D. F. (1997). Success criteria and adaptive management for a large-scale wetland restoration project. *Wetlands Ecol. & Manag.*, 4, 111-127.

Weinstein, M. P., Teal, J. M., Balletto, J. H. & Strait, K. A. (2001). Restoration principles emerging from one of the world's largest tidal marsh restoration projects. *Wetlands & Ecol. Manag.*, 9, 387-407.

Weinstein, M. P. (2007). Linking restoration ecology and ecological restoration in estuarine landscapes. *Estuaries & Coasts*, 30, 365-370.

Wiegert, R. G. & Freeman, B. J. (1990). Tidal salt marshes of the Southeast Atlantic Coast: a community profile. *Biological Report 85(7.29). Washington, D.C.:US Fish and Wildlife Service.*

Young T. P., Petersen, D. A., & Clary, J. J. (2005). The ecology of restoration: historical links, emerging issues and unexplored realms. *Ecology Letters*, 8, 662-673.

Zedler, J.B. 1993. Canopy architecture of natural and planted cordgrass marshes: selecting habitat evaluation criteria. *Ecological Applications*. 3: 123-138.

Zedler, J. B. & Callaway, J. C. (1999). Tracking wetland restoration: Do mitigation sites follow desired trajectories? *Rest. Ecol.*, 7, 69-73.

Zedler, J. B. & Callaway, J. C. (2003). Adaptive restoration: A strategic approach for integrating research into restoration projects. In D. J. Rapport, W. L. Lasley, D. E. Rolston, N. O. Nielsen, C. O. Qualset, & A. B. Damania (Eds.), *Managing for Healthy Ecosystems* (pp. 167-174). Boca Raton, FL: Lewis Publishers.

Zheng, L., Stevenson, R. J. & Craft, C. (2004). Changes in benthic algal attributes during salt marsh restoration. *Wetlands,* 24, 309-323.

In: Ecological Restoration
Editors: George H. Pardue and Thomas K. Olvera

ISBN 978-1-60741-013-3
© 2009 Nova Science Publishers, Inc.

Chapter 5

RESTORATION OF CHANNELIZED FLUVIAL SYSTEMS AND THEIR FLOODPLAINS

Aaron R. Pierce

Department of Biological Sciences, Nicholls State University,
Thibodaux, LA 70310, USA

ABSTRACT

Anthropogenic alteration of fluvial systems has occurred extensively throughout the world. Many of these alterations, including levees, dams, and channelization, disrupt functional processes of these systems and their associated floodplains. Most prominent among these disruptions is the connectivity of the river channel to the floodplain that is critical for functions of productivity, decomposition, nutrient cycling, and the biota adapted to seasonal wet/dry periods. In addition, the effects of activities such as channelization can be exacerbated by regional geology and surrounding land-use activities, resulting in dramatic geomorphic readjustments. Currently, there is considerable interest in restoring fluvial systems and their floodplains because of their valuable functions to both ecosystems and society, including water quality enhancement, flood control, erosion control, timber production, recreational opportunities, and wildlife habitat. This chapter focuses on fluvial and floodplain processes, the effects of anthropogenic alterations these functional processes, and discusses restoration guidelines for these systems.

INTRODUCTION

Alluvial floodplains are associated with low-gradient rivers and experience periodic inundation. In the United States, these floodplain systems are most abundant in the southeast including regions of the Piedmont, Coastal Plain, and Lower Mississippi Alluvial Valley (Hunt 1967). Floodplains maintain biologically diverse and remarkably productive ecosystems that are adapted to fluctuating water levels (Odum 1969) defined and sustained by a natural hydrology of alternating dry and wet periods (Wharton et al. 1982).

Floodplains provide numerous valuable functions to both society and nature. These benefits include water quality enhancement, flood control, erosion control, timber production, and wildlife habitat. However, since the 1700's fluvial systems and their associated floodplains in the United States have been destroyed and degraded (Hefner and Brown 1985). Human alteration of hydrologic regimes including dams, impoundments, regulated water flow, wetland drainage, habitat fragmentation and groundwater extraction have reduced wetlands, including floodplain ecosystems, in the United States by 50% (Pringle 2000). Palustrine vegetated wetlands, including floodplain forests, have experienced the greatest net loss of all wetland types (Hefner and Brown 1985, Wilen and Frayer 1990). Human alterations may not only direct reduce the amount of wetland and floodplain habitat but also alter fluvial geomorphic processes and therefore the functioning of the floodplain system (Figure 1).

Figure 1. Photograph demonstrating the reduction of floodplain habitat in response to channelization and agriculture.

The Lower Mississippi River Alluvial Valley (LMAV) originally contained over 10 million ha of floodplain forests, which has been reduced to a highly degraded 2.8 million ha (Hefner and Brown 1985). Nearly all floodplain forest loss has been attributed to clearing, draining, and other hydrologic alterations for agriculture (Wilen and Frayer 1990). The loss and degradation of floodplain forests have had negative effects on wildlife communities including Neotropical migratory songbirds (Hunter et al. 1993), waterfowl (Heitmeyer and Fredrickson 1981), and fish (Hoover and Kilgore 1997, Risotto and Turner 1985). A clear understanding of the processes involved in floodplain ecosystems is necessary to determine the impacts of anthropogenic disturbances on these valuable systems and for their future restoration and management.

FLOODPLAIN FORMATION

The alluvial floodplain is a broad flat stretch of land occurring as a result of the lateral outward and downward progression of the stream combined with erosion and deposition that occurs during overbank flooding (Leopold et al. 1992). The energy of flowing water that facilitates erosion and deposition of sediment is responsible for the morphology of the floodplain and establishes the complex relationship between the hydroperiod and floodplain communities. The rivers and the sediment they carry are responsible for the origin, character, and maintenance of the floodplain and its plant and wildlife communities.

In the Coastal Plain Physiographic Province of North America, which contains the greatest extent of alluvial floodplains (Hunt 1967), high energy flows from steep headwater areas encounter highly erodible alluvium of the Coastal Plain adding sediment to the river flow (Wharton et al. 1982). Sediment is transported downstream in the form of suspended sediment (fine-grain particles) and bedload (coarse-grain particles). As the sediment load increases the velocity of the flow decreases. This dissipation of energy results in the deposition of alluvium that is reworked by fluvial geomorphic processes into meanders. The energy of the flowing river is decreased by lateral meandering that widens the floodplain (Knighton 1998). Rivers adjust their slopes by meandering until a steady state is reached with a balance of sediment load, water velocity, and volume. As floodplains become wider, more sediment is deposited from overbank flooding. Broad, flat floodplains are produced through these processes (Figure 2).

Figure 2. Photograph illustrating a broad, flat floodplain; including meanders and oxbow lakes that are produced through fluvial geomorphic processes.

SEDIMENT

Variables that determine the behavior of water and sediment include: climate, geology, soils, land use and vegetation (Wharton et al. 1982). These variables are highly interdependent and control the dominant depositional and erosional processes on floodplains. These processes include (1) sheet or gully erosion, (2) point bar deposition, and (3) overbank deposition.

Sediment in floodplain systems originates from the continuous erosion of the landscape through time. Sands, silts, and clays are transported into streams by sheet-wash or gully erosion. Land use activities in the uplands influence the quantities of sediment entering the alluvial system. Before the 1900's the sediment inputs in the southeastern river systems were minimal (Wharton et al. 1982). Erosion from forested uplands is estimated to be relatively small, 2.5 cm/ha per 16,000 years (Soil Conservation Service 1977). In contrast, loss of sediment from croplands can be more than 87,000 kg/ha/year as reported for the Obion-Forked Deer River in TN (Soil Conservation Service 1977).

Gullies are unstable landforms that are dynamic and constantly changing as part of a developing drainage network (Figure 3). Gullies are "relatively deep, recently formed eroding channels that form on valley sides and valley floors where no well-defined channel previously existed" (Schumm et al. 1984). Gully erosion typically occurs in steep ephemeral stream valleys that are actively eroding. Gullies are also known to develop rapidly as a result of agricultural practices by acquiring adjacent rills and developing a branching network of ephemeral tributaries (Bloom 1978).

Figure 3. Photograph of a massive gully in an agricultural setting that contributes sediment into a fluvial system (upper right corner).

Once in the stream system, sands are transported as bedload moving in wave-like motion along the channel bed. Silts and clays are transported as suspended sediment within the water column. Movement of the sediment through the system is not continuous but is a result of discrete climatic events. The largest portion of the total sediment load is usually carried by one or two high flows per year (Knighton 1998).

Deposition of the sediment load occurs through point bar deposition and vertical accretion on the floodplain. Point bar deposition occurs on the convex banks of river meanders by lateral accretion. This process is coupled with erosion of the opposite concave banks of the meanders resulting in the bulk of sediment being stored within the floodplain (Leopold and Wolman 1957, Hupp 2000). Deposition in the floodplain (vertical accretion) occurs during flood events where water flows are forced outward into the floodplain causing deposition and forming the primary floodplain soils. Floodplain deposition is highly variable depending on frequency of flooding, distance to channel, sediment load, sediment texture, water velocity, and floodplain morphology.

During overbank flooding, water velocities slow resulting in coarse sediment deposition (sand) near the channel thus developing natural levees (Happ et al. 1940 Pierce and King 2008). As flows penetrate further into the floodplain additional velocity is lost due to the roughness of the floodplain causing slower deposition rates of fine sediments in flats and backswamps, where water is ponded for relatively long periods (Happ et al. 1940, Hodges 1997, Hupp 2000). High velocity flows through sheet flow and scour channels can also transport coarse bedload deep into the floodplain (Wolman and Leopold 1957, Pierce and King 2008). Overbank deposits may also be layered, with coarse sand layers representing the rise to maximum stage of a flood event and fine-grained layers representing silt and clay deposits as flows retreat (Allen 1965).

Average floodplain deposition ranges between 0.3 m and 0.6 m in 200 to 400 years (Wolman and Leopold 1957, Hupp and Bazemore 1993, Kleiss 1996, Heimann and Roell 2000). Overbank flooding can also produce local erosion in forested floodplains. Local erosion and deposition occurring within the floodplain create a spatial mosaic of aggrading and degrading surfaces on top of material already deposited.

HYDROLOGY

The hydrological regime is the ultimate driving force in floodplain communities and is responsible for creating and maintaining these systems. Hydrologic energy is crucial for the transportation and redistribution of sediment within the system. In the southeastern United States, dramatic fluctuations in water levels are the result of high flow periods due to winter-spring rains (Figure 4a) and low flows attributed to dry summer and fall months with high evapotranspiration (Figure 4b) (Wharton and Brinson 1979). High runoff during the winter and spring months usually overflows the floodplain features. This high water level is sometimes sustained by the cumulative effect of many tributaries and isolated rainfall. Infrequent rain events during late summer and fall coupled with high evapotranspiration can result in near zero discharge. The duration and frequency of flooding in alluvial floodplains is dependent upon several factors including: rainfall, size and slope of the watershed, soils, and elevation (Hupp 2000). These factors also explain the various forest communities located in the floodplain.

A)

B)

Figure 4. (A) Photograph of an inundated floodplain during an overbank flooding event occurring during winter. (B) Photograph illustrating the productivity of floodplain systems during low flow periods occurring in summer.

FLOODPLAIN FEATURES

Although floodplains have little topographic relief, the continuous erosion and deposition of sediment creates floodplains with diverse micro-topographic (geomorphic features) that define distinct ecological zones within floodplains. Small differences in elevation can result in unique site conditions that affect the stand composition (Hodges and Switzer 1979). The following section describes the various floodplain features that create the ecological zones within the floodplain.

Channels

River channels are responsible for the creation and maintenance of floodplains. Long-term flow patterns are responsible for stream channel morphology (width, depth, slope, etc.). High flow events are most important in creating channels and their morphology, such as meanders. Sinuous meanders are a basic feature of unaltered alluvial systems that enables rivers to adjust to the slope of the floodplain (Leopold and Langbein 1966). Meanders allow rivers to stabilize water velocity by balancing the distance the water travels with the steepness of the slope. Once the meander is formed, the flow continues straight until the velocity is slowed enough that gravity can turn the water downstream again and the process is repeated. The channel slope also naturally adjusts to the velocity of the flow and the energy needed to transport sediment through the system. As the gradient increases, the stream must degrade or develop meanders to accommodate the increased velocity (Leopold and Langbein 1966).

During overbank flooding events, the floodplain becomes part of the channel. This allows the water to flow directly downslope with only the floodplain surface and forest vegetation to slow the velocity. This allows flow to transport sediment into floodplain before being deposited. During flood events, anastomosing channels may also form where there is a change in slope or some impediment that causes excessive coarse sediment to be deposited in the main channel. Multiple channels are then formed around the plug of sediment (valley plug) (Pierce and King 2008). The anastomosing channels may traverse into the floodplain and rejoin downstream of the plug.

Natural Levees

As overbank floodwaters spread into the floodplain, flow velocity decreases due to vegetation and the roughness of the floodplain (Happ et al. 1940, Johnston et al. 1984, Hodges 1997). The decreased flow velocity enables deposition of sands and silt adjacent to the channel forming natural levees along the channel (Hupp 2000). Natural levees are usually more developed on concave stream banks than convex and average between 30 m and 100 m wide (Wharton et al. 1982). If there is a break in the natural levee, sediment can spread out into the floodplain creating a fan-shaped crevasse-splay (Allen 1965, Bloom 1978, Knighton 1998).

Floodbasins and Backswamps

Floodbasins refer to underfitted floodplains where the channel and its meanders only occupy a small portion of the floodplain width. Floodbasins were created under significantly higher flows than the current flows. Flats and backswamps apply to systems where the floodplain is not underfit and there is a shallow depression between the natural levee and the valley wall. Sediment deposition that occurs in both of these features comprise of fine silt and clay materials (Hupp and Morris 1990, Kleiss 1996). Areas that remain flooded or that have water tables near the surface are known as backswamps; these areas experience little deposition.

Ridge and Swale Topography

As erosion occurs on concave banks of meanders and point bar deposition occurs on convex banks, ridge and swale topography is formed. Small ridges form on the convex banks due to deposition forming a temporary natural levee. A series of ridges form, with intervening depressions called swales, as the river moves laterally and downstream (Wharton et al. 1982, Hupp 2000). Vegetation rapidly colonizes and stabilizes each ridge that facilitates further deposition. Ridges are mostly comprised of sands while silts and clays become deposited in swales (Wharton et al. 1982).

Scour Channels, Hummocks, and Mini-Basins

These features all contribute to the microtopography of the floodplain. Although these features may only produce slight elevational changes, they can have significant impact on plant species distributions (Wharton et al. 1982, Hodges 1997).

Scour channels are formed during high flow events that cut small channels through the floodplain. These channels are typically dry during most of the year except during periodic flood events. Scour channels contain a high percent of sand as does the adjacent floodplain because these areas usually transport sheet flow that can carry substantial sand bedload across the floodplain (Wharton et al. 1982).

Hummocks are islands created after extended periods of erosion by scour channels. These islands are usually vegetated which help to prevent additional erosion of these areas. Often hummocks are occupied by tree species commonly found at higher elevations, but can also contain more flood tolerant species like cypress and tupelo (Wharton et al. 1982).

Minibasins are shallow depressions distributed throughout the floodplain, often occurring between the bases of trees. They are typically filled with rainwater and collect detritus that is quickly decomposed due to the frequent fluctuations between wet and dry conditions (Wharton et al. 1982). Thus, minibasins can contribute a great deal to the nutrient cycling of the floodplain.

FLOODPLAIN FORESTS

Composition and Structure

Floodplain forests support recognizably distinct assemblages of plants and animals that are associated with particular landforms, soils, and hydrologic regimes (Hodges and Switzer 1979, Mitsch and Gosselink 2007). Flood frequency, depth, timing and duration are primary determinants of floodplain species composition (Hodges 1997, Hall and Harcombe 1998). To a lesser extent, composition is also influenced by an array of factors within the floodplain including: groundwater levels, soils, micro-topography, light intensity, and human activities (Clark and Benforado 1981). Due to the relationship with hydrology, elevation and deposition rates contribute to major differences in site conditions, which determine the plant communities (Hodges 1997). The distribution of floodplain tree species occurs along a hydrologic gradient perpendicular to the river channel (Wharton et al. 1982, Kozlowski 2002). This gradient is related to the micro-topography and geomorphic features within the floodplain.

The hydrologic gradient within the floodplain, also known as a moisture or anaerobic gradient, is the primary factor influencing plant species distribution within the floodplain (Wharton et al. 1982). The distributions are determined by three aspects of the anaerobic gradient: (1) intense selective force of anaerobic conditions provided by the hydroperiod of the floodplain, (2) the spatial and temporal variability of the anaerobic gradient across the floodplain due to micro-topography, soils, and the hydroperiod, and (3) the tolerances of plant species to the anaerobic gradient (Wharton et al. 1982).

Inundation creates several anaerobic stresses or selective pressures on plant species inhabiting these sites due to physical and chemical changes in the saturated substrate. These changes include: (1) depletion of available oxygen, (2) shifts in soil pH, (3) buildup of toxic compounds in the plant, and (4) changes in nutrient availabilities (Wharton et al. 1982, Kozlowski 2002). Floodplain tree species have evolved adaptations, of varying degrees, to deal with these stresses. Individual species tolerances to anaerobic stresses determine their distribution within the floodplain.

Flood tolerant species have developed both physical and metabolic adaptations to deal with stresses from inundation. Physical adaptations include adventitious and secondary root development that increases the uptake of oxygen, facilitates the movement of nutrients and gas exchange, and increases anaerobic respiration (Wharton et al. 1982). Other physical adaptations include pneumatophores (knees), buttress formation, and wide, shallow root formations all of which provide support, increased oxygen efficiency, reduced nutrient loss, and protection from erosion (Wharton et al. 1982). Metabolic adaptations include (1) shifts in end products to prevent ethanol toxicity, (2) oxidization of the rhizosphere to prevent root deterioration, and (3) substitution of nitrate for oxygen in cellular reactions when oxygen is depleted and nitrate is abundant (Wharton et al. 1982, Kozlowski 2002). The influence of timing, depth, duration, and frequency of flooding on plant species composition, distribution and survival is most critical during the growing season. Floods during the dormant period have a modest effect on physiology and survival of plant species (Wharton et al. 1982).

Although plant communities in floodplain systems are extremely diverse, broad generalizations of tree species community types or associations have been identified that correspond to the different geomorphic features within the floodplain (Hodges 1997, Conner and Sharitz 2005). The first community is located along the stream edge or bar. This area is characterized by frequent flooding and high rates of coarse sediment deposition. Black willow (*Salix nigra*), cottonwood (*Populus* spp.), and river birch (*Betula nigra*) are typically dominant species in this community (Wharton et al. 1982, Hodges 1997, Conner and Sharitz 2005). These species tend to have large seed crops, rapid germination and rapid growth rates that allow them to thrive in these areas.

The front or natural levee tends to be of slightly higher elevation than backswamps or fronts, is typically composed of coarse-grained sediments, and is seasonally and temporally flooded at least once every three years (Mitsch and Gosselink 2007). The dominant tree species on natural levees are sycamore (*Platanus occidentalis*), sweetgum (*Liquidambar styraciflua*), and elm (*Ulmus* spp.) (Wharton et al. 1982, Hodges 1997, Conner and Sharitz 2005).

Backswamps or first terrace flats are usually temporally flooded at least once every two years and are typically composed of heavy clay soils (Mitsch and Gosselink 2007). Backswamps are colonized by overcup oak (*Quercus lyrata*), water hickory (*Carya aquatica*), green ash (*Fraxinus pennsylvanica*), sugarberry (*Celtis laevigata*), and elm (Wharton et al. 1982, Hodges 1997, Conner and Sharitz 2005). Sloughs or swamps within the first terrace are semi-permanently flooded, and typically contain baldcypress (*Taxodium distichum*) and water tupelo (*Nyssa aquatica*) (Wharton et al. 1982, Hodges 1997, Conner and Sharitz 2005). Ridges are higher elevational areas within the first terrace that are dominated by sweetgum, water oak (*Q. nigra*), willow oak (*Q. phellos*), hickories (*Carya* spp.), blackgum (*N. sylvatica*), and elm (Wharton et al. 1982, Hodges 1997, Conner and Sharitz 2005).

The second terrace is seasonally and temporarily flooded once every four years (Mitsch and Gosselink 2007). This area typically consists of sweetgum, willow oak, water oak, cherrybark oak (*Q. falcata* var. *pagodifolia*), swamp chestnut oak (*Q. michauxii*), hickory, elm, and blackgum (Wharton et al. 1982, Hodges 1997, Conner and Sharitz 2005). Ridges within the second terrace can be colonized by upland species including: oak, hickory, ash, elm, loblolly pine (*Pinus taeda*), and yellow polar (*Liriodendron tulipifera*) (Wharton et al. 1982, Hodges 1997, Conner and Sharitz 2005).

Floodplain Forest Succession

Plant succession is a driving force for plant community structure and composition. Plant succession can be defined as the "non-seasonal, directional, and continuous pattern of colonization and extinction on a site by species' populations" (Begon et al. 1990). Plant community level changes are therefore the result of individual and population level changes. Changes can occur due to disturbance, dispersal of new species or individuals, competition between individuals and species, and modification of the site by individual interactions or external factors (Barnes et al. 1998). Changes can occur in different forms and at different degrees of intensity, all contributing to the succession of the plant community.

Plant community succession is driven by several processes. Understanding these processes is essential for wetland conservation and management. One of the most important processes driving succession in wetland habitats is disturbance. Disturbance is any "discrete event in time that alters the ecosystem, community, or population structure and changes resources, substrate availability, or the physical environment" (Pickett and White 1985). Disturbances can come in many forms, frequency, intensity, and size; all of which can impact the plant community in different ways depending on the species involved.

Floodplain forest succession involves both autogenic and allogenic processes that direct succession (Hodges 1997). Autogenic processes are plant driven processes by which the present plant community changes the environment making it more suitable for colonization of other species (Barnes et al. 1998, Mitsch and Gosselink 2007). Allogenic processes are external factors that change the environment such as disturbance and other physical factors (Barnes et al. 1998, Mitsch and Gosselink 2007). In floodplain systems, the important allogenic processes directing succession are hydrology and sediment deposition, which are highly interdependent processes.

Three general modes of succession in floodplain systems have been identified by Hodges (1997). The first pattern of succession occurs on permanently flooded sites that receive little sediment deposition. In this situation, succession is halted, with no changes in composition for centuries unless a disturbance occurs. Floodplain sites that are typically affected by this type of succession consist of baldcypress-tupelo swamps.

The second pattern of succession occurs at low elevation sites that experience longer hydroperiods and slow rates of deposition consisting of fine sediments. These areas are typically colonized first by black willow for approximately 30 years. Areas that experience little or no deposition may then be colonized by swamp-privet (*Forestiera acuminata*), water elm (*Planera aquatica*), and buttonbush (*Cephalathus occidentalis*). If deposition continues to be minor, baldcypress may replace this community. If there is adequate deposition, then the area may be dominated by an overcup oak – water hickory type. However, if deposition is still slow but sufficient the willow stand may go directly to an overcup oak – water hickory type, skipping the swamp-privet, water elm, and buttonbush phase. If the deposition is rapid but texture remains fine, the willow stand will develop into an elm-ash-sugarberry type. If the rapid deposition consists of coarse sands, then an association of boxelder (*Acer negundo*), silver maple (*A. saccharinum*), and sugarberry will develop.

Forests under the influence of the second pattern of succession may develop through different phases and repeat phases due to disturbances, but all lead to elm-ash-sugarberry type. This type will replace itself and persist unless there is considerable change at the site. Small disturbances may result in a shift toward an oak-sweetgum forest usually shifting back after time to elm-ash-sugarberry type. A lack of disturbance at the site will over time shift the elm-ash-sugarberry type to a climax association of oak-hickory. The time period necessary for the oak-hickory association to develop is 600 years.

The third pattern of succession occurs on higher elevations in the floodplain including ridges and natural levees. First species to colonize these sites consist of cottonwood and black willow. This association is then replaced by boxelder, sugarberry, hickory, and sliver maple if the break up of cottonwood is gradual. Over time this forest will develop into the elm-ash-sugarberry type. However, if the break up of cottonwood is rapid, then sycamore, elm, ash, sweetgum, willow oak, and water oak will colonize the site. This association will then progress to the elm-ash-sugarberry type unless a disturbance rapidly opens the area, then an oak-sweetgum forest will develop. After time, this type will also convert to the elm-ash-sugarberry type. When disturbances of flooding and deposition diminish the climax type of oak-hickory will develop. The climax oak-hickory association is initiated with cherrybark oak and swamp chestnut oak and will take approximately 200 years to fully develop into the climax association (Shelford 1954).

The floodplain community level changes, described by Hodges (1997), are a result of differential survival of overstory trees, seedlings, and germination rates of seeds under prevailing hydrologic and sediment deposition conditions. The hydrological regime including timing, depth, frequency and duration of inundation has been shown to influence tree growth and mortality (Keeland and Sharitz 1997). The hydrologic regime is also the major factor affecting seedling establishment, growth, and survival (Streng et al. 1989, Johnson 2000). Germination potential of floodplain tree species under different flooding regimes has received little study. Past germination studies have shown little effect of flood duration or depth on germination rates but sediment deposition regimes can have a significant effect on tree growth, mortality, establishment, and germination (Walls et al. 2005, Pierce and King 2007a).

ALTERATIONS TO FLUVIAL SYSTEMS

Unfortunately, most fluvial systems are unable to function as described above because of human alterations to fluvial systems, such as dam construction, channelization, and land use practices. These alterations are not only widespread throughout the United States but also the world (Sharitz and Mitsch 1993, Mitsch and Gosselink 2007). Human alterations to the fluvial system not only disrupt hydrological processes but also alter floodplain ecosystem services, such as sediment trapping, nutrient cycling, plant community dynamics, and fish and wildlife communities.

Dam construction has altered fluvial processes of most of the world's largest rivers (Nilsson et al. 2005). This type of alteration has the most significant impacts downstream of the dam (Petts and Gurnell 2005) and typically includes reductions in peak flood stages, frequency of floods, and duration of overbank floods. Regulation of the flow regimes in these downstream reaches concentrates flows on mid- and low- bank slopes and contributes to decreased sediment transport downstream of the dam (Williams and Wolman 1984, Church 1995, Brandt 2000). As a result of dams, the regulated and confined flows may contribute to channel and bank erosion in the upper downstream reaches from the dam (Williams and Wolman 1984). Concurrently, lower reaches may accumulate large amounts of sediment resulting in a homogenized and elevated floodplain that can adversely impact floodplain biodiversity (Hupp 1999, Hupp et al. *In press*). Flow alterations produced by dam construction not only reduce high-magnitude flooding and create stable flows, but also induce geomorphic adjustments that influence floodplain productivity and cause ecosystem degradation (Richter et al. 1996, Hupp et al. *In press*).

Channelization of fluvial systems is another common human alteration that has occurred extensively worldwide. Although there is no current worldwide or national assessment of the number of streams that have been channelized, this type of alteration is thought to be more extensive than damming (National Research Council 1992). Throughout the southeastern United States, channelization has been a common approach to reduce flooding, mainly for agricultural purposes (Shankman 1993). Channelization involves widening and deepening a stream channel to increase channel capacity, shortening stream length, and increasing the stream gradient to increase water transport (Figure 5).

Figure 5. Example of a channelized stream in western Tennessee.

The increased channel capacity and increased transport efficiency of channelized reaches cause channels to be disconnected from the adjacent floodplain (Kroes and Hupp *In press*). Channelization has been shown to reduce flooding in upstream reaches of a system, while causing lower reaches to experience increased peak flood stage and flood frequency (Shankman and Pugh 1992). Channelization and dredging of stream channels has also been shown to lower the water table levels in the floodplain (Tucci and Hileman 1992). Channelization can initiate further degradation of the stream channel through head-cutting and channel erosion resulting in bank failure (Robbins and Simon 1983, Simon and Hupp 1987, Simon 1994). Channelization increases stream power, thereby facilitating sediment transport and causing lower reaches to accumulate large amounts of sediment because decreased stream gradients and channel obstructions (i.e. debris jams, Figure 6) reduce flow velocities resulting in deposition (Schumm et al. 1984, Simon and Hupp 1987). This process causes a widening of the stream channel as it fills in with sediment which has been identified as a critical stage of the geomorphic recovery from channelization (Schumm et al. 1984, Simon and Hupp 1987). However, in some cases, mainly due to the geology of the region and land use practices, severe accumulation occurs and the channels become completely blocked with sediment (Figure 7), known as valley plugs (Happ et al. 1940, Diehl 2000, Pierce and King 2007b, 2008).

Figure 6. Example of a debris jam that contributed to the formation of a valley plug on Hickory Creek in western Tennessee.

Figure 7. Example of a channelized stream that has filled in with coarse sand and formed a valley plug on Jeffers Creek in western Tennessee.

The fluvial geomorphic effects of channelization can also disrupt functional processes of floodplain systems (Schumm et al. 1984, Simon and Hupp 1992, Brierley and Murn 1997, Brooks and Brierley 2004, Pierce and King 2008). Channelization has been shown to disrupt system functions in the following ways: altered surface and sub-surface hydrology (Shankman and Pugh 1992, Tucci and Hileman 1992), altered sedimentation rates (Happ et al. 1940, Hupp and Bazemore 1993, Pierce and King 2007b, 2008), reduced lateral channel migration responsible for creating sloughs and oxbow lakes (Shankman 1993), loss of aquatic habitat (Hohensinner et al. 2004), reduced growth and increased mortality of floodplain tree species (USDA 1986), loss of plant species diversity (Miller 1990), changes in plant species composition (Oswalt and King 2005, Pierce 2005), and negative effects on fish and wildlife communities (Heitmeyer and Fredrickson 1981, Risotto and Turner 1985, Hunter et al. 1993, Hoover and Kilgore 1997).

CASE STUDY: WESTERN TENNESSEE

Geology

The western Tennessee region is bordered by the Tennessee River on the east and the Mississippi River on the west. This region encompasses both the Gulf Coastal Plain and the Lower Mississippi Alluvial Valley. The parent material of this region is unconsolidated coarse sand deposited mainly during the Quaternary age (Saucier 1994). There is no bedrock control in this area for the base level of streams flowing into the Mississippi River. Thus, streams freely adjust their profiles to recover from disturbances such as dredging and channelization (Simon 1994).

The unconsolidated alluvial sands of western Tennessee are mostly covered by a thin layer of windblown loess deposits (silt and clay). These deposits range in depth from 1 to 30 m and are cut through by most of the tributary systems of this region (Saucier 1994, Simon 1994). Some tributaries of the Upper Hatchie River Watershed originate on block clay deposits known as Porters Creek Clay of the Midway Group, deposited during the Tertiary age (Miller et al. 1966). However, the main river systems of this region, including: the Obion River, the Forked Deer, and Hatchie River cut through the alluvial sands deposited during the Quaternary age (Saucier 1994).

Both the loess and alluvial sand deposits are highly erodible. Most of the erosion in western Tennessee is thought to occur in the loess-capped uplands. This erosion consists of both the loess cap and the alluvial sands beneath. The past 150 years of erosion in the uplands of western Tennessee exceeds the erosion that occurred in the last several thousand or tens of thousands of years (Saucier 1994).

Land-Use and Alterations

The western Tennessee region was rapidly colonized in the early 1800's. Upland areas in the region were quickly cleared for the timber value and replaced with agriculture fields of corn, cotton and tobacco (Wilder 1998). Before deforestation occurred, the rivers within the region were described as flowing at "constant good depths" (Ashley 1910). However, clearing of the upland areas resulted in erosion and gullying of the loess and sandy soils (Simon 1994). Before settlement, sediment deposition rates in western Tennessee ranged from 0.02 to 0.09 cm/year with post-settlement rates increasing to 3 cm/year (Wolfe and Diehl 1993). The increased erosion after settlement lead to deposition in the floodplains and stream channels causing a decrease in channel flood capacity. Streams were stifled with sediment and debris leading to frequent and prolonged flooding in the bottomlands (Mogan and McCrory 1910). Stream channel alteration was proposed to alleviate the flooding problems (Hidinger and Morgan 1912).

By 1926, most streams in western Tennessee had been channelized, resulting in over 132 km of stream alteration (Speer et al. 1965, Simon and Robbins 1987). The Hatchie River main stem was the exception to the widespread channelization. However, most of the tributaries of the Hatchie River were channelized (Simon and Hupp 1992, Simon 1994). Clearing and snagging of altered systems was necessary due to sediment aggradation and debris accumulation. Altered systems required continued maintenance from the 1930s through the 1950s (Simon 1994).

By 1970, the sediment build-up within the tributary floodplains was so great that the U.S. Department of Agriculture (1970) initiated a channelization project called the West Tennessee Tributaries Project (WTTP). Although only 35% of the project was completed, 128 km of channel alterations occurred because of the project (Robbins and Simon 1983). Channelization in western Tennessee resulted in streams being shortened by 44%, lowered by 170%, and steepened by 600% (Simon and Hupp 1992). The Obion-Forked Deer River system suffered considerably from alterations to its hydrology (Simon 1994). Impacts to the Hatchie River were restricted to its tributary system with 33 of its 36 major tributaries being channelized (USDA 1986).

The channelization projects in western Tennessee lowered the bed level in stretches of the channels by as much as 5 meters (Simon and Hupp 1987, Simon 1994). Transition slopes were constructed to offset differences in bed elevations at the junction of the modified and natural channel reaches (Robbins and Simon 1983). Transition slopes were steeper than both modified and natural channel reaches resulting in headcutting and degradation of upstream reaches. This degradation moved upstream at a rate of 2.6 km/year on the South Fork Forked Deer River and resulted in approximately 2.6 m of incision from 1966 to 1967 (Simon 1994). Downstream stretches accumulated the eroded material from upstream reaches. Deposition rates on downstream stretches of the Obion River ranged between 0.03 to 0.12 m/year, while the South Fork Forked Deer River filled in with 2.2 m of sediment over a 12-year period (Simon 1994). Deposition events of 0.61 m were also observed in the Reelfoot lake area (Shelford 1954).

By 1971, the completed channelization projects had directly and indirectly reduced the floodplain forest habitat along the affected reaches by 60% (Barstow 1971). Systems not directly impacted by channelization projects experienced losses of floodplain forest due to clearing of the floodplain and construction of drainage ditches by individual landowners (Barstow 1971). During this period of drastic hydrologic alteration, floodplain forests in western Tennessee were reduced from 404,000 ha in 1940 to 291,00 ha in 1970 (Turner et al. 1981). Land use changes in western Tennessee during the 1800s to the mid-1900s, were mainly driven by agricultural needs or desires.

Current Problems

Historically, the high meandering rate and low gradient of the rivers in western Tennessee did not allow for transport of large quantities of sediment. However, channelization of all rivers, except the Hatchie River, greatly increased their stream power and has led to dramatic geomorphic changes (Diehl 2000, Oswalt and King 2005, Pierce and King 2007b, 2008). Although the Hatchie River has not been channelized, 92% of its tributaries were channelized, forming a transport network for sediment into the Hatchie River. As a result, an estimated 580 million kilograms of sediment is accumulating in the Hatchie River every year (USDA Conservation Service 1986).

Channelization has contributed to the problem in two main ways. First, it alters the channel morphology by shortening, straightening and increasing the slope of the channel. As previously discussed (Robbins and Simon 1983, Simon and Hupp 1987, Simon 1994) such alterations cause a degradation of the stream channel leading to channel erosion. Several studies on sediment dynamics in western Tennessee suggest that channelization leads to bed-level lowering and stream degradation (Hupp and Simon 1986, Diehl 2000). Currently, channel erosion and bank failure (Figure 8) are thought to be contributing the majority of sediment into these channelized fluvial systems (Boulton 2005).

Figure 8. Photograph of a channelized stream that is experiencing sediment inputs from bank failure.

The second major impact of channelization on these systems is the facilitation of sediment transport. The channel alterations produced by channelization combine to increase stream velocity and stream power (Gilvear and Bravard 1996). The increased stream power facilitates sediment transport downstream through suspended fine sediments and coarse bedload movement while also increasing channel bank and bed erosion (Happ et al. 1940).

Bedload deposition takes place when the stream power is reduced below the transport threshold. This typically occurs in altered systems at woody debris jams, where the steep slopes of tributaries with high sand bedloads meet with lower slopes of downstream sections, and at the confluence of tributaries and the main channel (Happ et al. 1940, Diehl 2000).

Changes in slope and debris jams reduce stream velocity, as a result, sediment deposition occurs in the main stem of the channel (Figure 7). Excessive bedload deposition is resulting in the formation of valley plugs or shoals throughout western Tennessee (Diehl 2000, Oswalt and King 2005, Pierce and King 2007b, 2008). Valley plugs are areas where the channel becomes completely filled with sediment, thus floodwater and sand bedload are forced out into the floodplain (Figure 9) (Happ et al. 1940, Pierce and King 2007b, 2008). The roughness of the floodplain reduces the water velocity causing additional deposition of sediment throughout the floodplain. This process spreads sediment throughout the floodplain as the stream braids-out from the main channel forming anastomosing streams throughout the floodplain (Happ et al. 1940, Diehl 2000, Pierce and King 2008). Sedimentation rates in floodplains adjacent to valley plugs were found to be almost 100-times greater than deposition rates in floodplains of unchannelized streams without valley plugs and consisted mostly of coarse sand (Pierce and King 2008).

Figure 9. Photograph of coarse sand deposited in the floodplain of Hickory Creek, western Tennessee, as a result of a downstream valley plug.

The processes discussed above are consistent with the numerous valley plugs located within the Hatchie River Watershed and other watersheds in the region (Diehl 2000, Oswalt and King 2005, Pierce and King 2007b, 2008). Since the channelization of the tributaries, the main channel of the Hatchie River has become shallower and flooding has increased (USDA 1986). This change has been concentrated near the mouths of several tributaries (Diehl 2000). Increased flooding in these areas is thought to inhibit growth and increase mortality of floodplain tree species and alter species composition of floodplain plant communities (Oswalt and King 2005, Pierce 2005).

Areas experiencing excessive deposition are buried by infertile sand, covering the productive silt-clay deposits. Excessive sedimentation can reduce the germination potential of typical floodplain tree species by burying seed sources in infertile sand (van der Valk et al. 1983, Walls et al. 2005, Pierce and King 2007a). Burial of established trees may also cause stress and mortality to mature tree species because of reduced nutrient availability and changes in groundwater levels. Rate and texture of sediment deposition may be important factors in determining the response of the vegetation (van der Valk et al. 1983, Walls et al. 2005, Pierce and King 2007a).

In addition to increasing sediment deposition, valley plugs and shoals also influence hydrology (Oswalt and King 2005, Pierce 2005). Increased overbank flooding events, as a result of channel deposition and reduced drainage capacity of the floodplain, can increase the frequency, depth and duration of flooding. Aggradation of sediment in stream channels can cause such dramatic increases in the water table that it can even be above some floodplain surfaces (Happ et al. 1940). Not only does the rise in water table contribute to increased swamping and ponding of the floodplain, but it can also inundate root systems of floodplain tree species throughout the growing season, causing reduced growth and mortality (King 1995, Keeland and Sharitz 1997, Johnson 2000). Prolonged inundation of floodplains in western Tennessee has caused total mortality of the original timber stands; allowing growth of only willow and other disturbance tolerant species (Happ et al. 1940, Oswalt and King 2005, Pierce 2005).

Overbank flooding and sedimentation are critical process in floodplain ecosystems that provide several benefits including replenished nutrients, fertile soil, and water recharging. However, channelization of fluvial systems, particularly in western Tennessee because of its unique geology and past land-use practices, severely affects fluvial-geomorphic processes at multiple spatial and temporal scales (Schumm et al. 1984, Simon and Hupp 1987) and thereby alters functional processes of floodplain systems including aquatic habitats, flood storage capacity of the stream, groundwater, sediment erosion and deposition rates and textures, overbank flooding regimes, plant community dynamics, and fish and wildlife communities.

Restoration Potential

Several restoration options have been proposed for the western Tennessee case study presented above, these include: (1) the "hands off" approach, where valley plugs are left in place to prevent sediment from moving downstream, (2) creating artificial valley plugs to prevent tributaries from transporting sediment further downstream, (3) restoring the fluvial portion of the tributaries, including restoration of the channels and gully regions, and (4) restoring both the fluvial systems and the floodplain.

The short-term deposition analysis indicates that valley plugs are protecting downstream sections from excessive sedimentation (Pierce 2005, Pierce and King 2008). This result is extremely important, as one of the restoration options under consideration is to leave valley plugs in place or add new ones to prevent large quantities of sediment from reaching downstream reaches. However, it is essential to note that this effect of downstream protection is only short-term. Field observations and pervious research (Happ et al. 1940) suggest that eventually the plug will be circumvented by the formation of new channels. This may result in formation of other plugs either upstream or downstream of the previous plug. For example, along a three mile stretch of Clover Creek, there were four former valley plugs in which the stream had created new channels through and around the valley plugs, resulting in a swamped floodplain similar to those, described by Miller (1990) and Oswalt and King (2005), along the Middle Fork-Forked Deer River. In the Hatchie River watershed, tributary floodplains that have been degraded by channelization and valley plugs have been reduced in economic value, mainly timber value, by $5,438 ha^{-1} (Wells 2004).

Another option that has been discussed is hydrologic restoration to reduce the transport capacity of the tributaries (Figure 10). This strategy may be effective if (1) there is also a stabilization of sediment sources, including gully erosion and channel erosion/bank failure, and (2) the restoration occurs over the entire system with particular attention to the stream gradient and channel depth at the confluence of other streams to prevent head-cutting. If these factors are included in the restoration project, then sediment input into the system would be stabilized and geomorphic readjustment caused by changes in stream power, flow velocity, and stream gradient may be minimized. However, this would only restore the fluvial system, which may benefit downstream reaches, but may not ensure restoration of the floodplain system. Although restoration of the fluvial system is a necessary first step in restoring the floodplain system, there are other processes involved that could prevent the establishment of typical floodplain tree species. Additional concerns include the loss of microtopography as a result of excessive deposition and seed availability/dispersal. Microtopography has a direct effect on flooding, which can determine the distribution of floodplain tree species (Hodges 1997). At valley plug sites, high deposition rates over a large extent of the floodplain may have reduced the microtopography of the floodplain. Research has shown that floodplain tree species respond differently to flood duration and deposition rates depending on life history characteristics (Walls et al. 2005, Pierce and King 2007a).

Seed availability may also be a limiting factor in the restoration of the floodplain. Studies (Oswalt and King 2005, Pierce 2005) indicated that common floodplain tree species including oaks, baldcypress, and tupelo are absent or rare at valley plug sites. Dispersal of seeds from upstream locations may also be minimal because of intensive agricultural practices that have reduced most of the upstream floodplain forests. Particular attention should also be given to the impact of channel restoration projects on water tables in the floodplain. It has been demonstrated that channel alterations can have a significant impact on water tables (Tucci and Hileman 1992, Pierce 2005), and therefore the site conditions in the floodplain for establishment of plant and wildlife communities.

Restoration of both the fluvial system and the floodplain has to consider the above-mentioned factors as well as others. Other factors included successful collaboration with landowners and overcoming communication difficulties and differences in goals. The cooperation of private landowners will be a critical factor in any restoration effort. Organizations involved in restoration activities have to be able to successfully communicate with landowners about the variability within these systems and the variety of restoration options open to them. Successful restoration will depend on their collaborative efforts.

Figure 10. Photograph of a Nature Conservancy restoration project on Richland Creek in western Tennessee. The project included instream limestone structures to encourage sediment deposition and bed-level adjustments, bank contouring, and planting several floodplain tree species.

RESTORATION GUIDELINES

In recent years, there has been considerable interest in restoring fluvial systems and their associated floodplains because of the valuable social and ecological functions that these systems provide, such as flood control, sediment and nutrient retention, recreational opportunities, timber production, and wildlife habitat. Despite our limited scientific database on restoration projects and poor predictive ability of restoration outcomes many restoration guidelines have been developed (Zedler 2000, Mitsch and Jorgensen 2004, Mitsch 2006). Some of these general guidelines are described below:

1. The project should have clearly stated and achievable goals with a mechanism in place for evaluation of the project's success. Projects should be designed to fulfill multiple goals, but need to be prioritized as primary and secondary objectives.

Selection of the appropriate restoration approach depends on the specific objectives of the project. Without clearly defined objectives, the restoration processes may be impeded by confusion and concerns among landowners, conservation organizations, state, and federal agencies. Unclear objectives can also prevent the appropriate evaluation of the project's success. Objectives may also be time sensitive; system processes must be repaired before appropriate vegetation can become established and herbaceous vegetation will respond more rapidly than woody vegetation (van der Valk 1998). Finally wildlife communities cannot be expected to be sustained by the system until vegetation and other habitat components have been established (Gawlik 2006). Therefore, temporal and successional aspects need to be incorporated into the objectives and evaluated with respect to those considerations.

2. Design the project for minimum maintenance and a general reliance on self-sustainability. This requires understanding and utilizing natural energy flows in the system and directing them for ecosystem function; particularly important is the hydrologic regime.

In order for a system to be self-sustaining particular attention must be given to restoration of the driving processes of the system (Zedler 2000). In fluvial systems, this requires restoration of the hydrological and geomorphic processes. Understanding the natural water flows in the system before alterations occurred is important for the restoration to be successful and sustainable (Hunt et al. 1999, Simenstad et al. 2006). This may require extensive engineering to return meanders to the fluvial system and ensure connectivity to the floodplain. In addition, accounting for bank instability and bed-level adjustments is also important to stabilize and restore geomorphic processes. Site selection is therefore another important consideration. If the hydrological and geomorphic processes are unlikely to be restored given the extent of alterations to the site, then the project may have little chance of being successful.

3. The project must be placed in the proper landscape context and position because project sites are typically part of a larger watershed and will influence as well as be influenced by the larger ecological setting.

Fluvial restoration projects are typically small, which helps ensure success of the project, but are also part of a larger watershed (Bedford 1999). Understanding how watershed scale processes and off-site alterations, such as other impoundments, may influence the functioning of the restoration site is (De Angelis et al. 1998, Mitsch and Day 2006, Simenstad et al. 2006). For example, restoring the middle reaches of a channelized system by incorporating stream meanders is unlikely to be successfully. High energy flows and high sediment loads will continue to impact the restored area from the channelized upstream reaches and head-cutting may proceed upstream from the channelized downstream reaches.

4. Do not overengineer the project especially with hard/rigid structures. Fluvial systems are always in a state of flux and to be self-sustaining they need to be able to change and evolve.

Rigid engineering structures can impede the natural state of flux in fluvial systems and therefore impact functional processes of the system. For example, hard structures, such as riprap, are sometimes used for bank stabilization. Although these structures may reduce channel erosion and widening they also prevent the establishment of vegetation and reduce aquatic habitat of the littoral zone. In addition, use of hard structures like detached breakwaters can divert hydrologic energy to unprotected areas and cause extensive erosion

(Piazza et al. 2005). However, in many restoration projects, water control structures are successfully utilized to ensure the appropriate hydrologic conditions (Mitsch et al. 1998). One downside to this type of technique is they require continued management and maintenance.

5. The success of all project goals can not be completed simultaneously. Ecosystem properties develop at different rates, therefore the project may need more time to develop all of the desired characteristics.

A critical factor for the improvement of ecological restoration is the evaluation of restoration outcomes. This requires continued monitoring of restoration projects and the appropriate evaluation of the success in achieving the project's objectives. Unfortunately, these aspects of restoration are often overlooked or not fully completed, in many cases due to lack of funding. However, for adaptive management to work and improve our restoration abilities; long-term monitoring and evaluation are essential. Long-term monitoring is particularly important when project goals maybe long-term, such as development of mature floodplain forests or sustaining forest bird communities.

6. Disturbance regimes should not be neglected. Disturbances such as flood pulses, sediment deposition, and fire can be a critical factor in the development of ecological functions and biodiversity.

In floodplain systems both plant and wildlife communities are adapted to periodic flooding which constitutes, in many cases, the main disturbance regime in these systems. Floodplains are able to maintain biologically diverse and remarkably productive ecosystems because of the hydrological and geomorphic disturbance regimes (Odum 1969, Hodges 1997, Ward and Trockner 2001). Without these disturbance regimes many of the valuable benefits that fluvial systems provide would not exist (Junk et al. 1989, Ward and Trockner 2001).

7. Hydrologically and biologically "open" systems can help return nutrients and propagules (plants, microbes, and animals) to the site. However, this can also introduce exotic species to the site that can disrupt species and community dynamics, thus monitoring is needed.

Utilizing natural energy flows and dispersal capabilities of plants and wildlife will help reduce maintenance costs and help establish ecosystem functions and processes necessary for a system to be self-sustaining (Mitsch and Gosselink 2007). However, this also requires an understanding of the surrounding landscape such as the availability and reliability of flow inputs and quality and plant and wildlife species that may potential invade the site. In some cases, this may include desired species as well as exotic or invasive species. Comprehensive monitoring may be needed to ensure the desired characteristics are established and maintained.

These principles are general guidelines for ecological restoration of fluvial systems based on pervious restoration work (Zedler 2000, Mitsch and Jorgensen 2004, Mitsch 2006). Historically, restoration work has focused mostly on structure of the system instead of function, which is necessary for the system to be successful in the long term and to be self-sustaining (Dahm et al. 1995, Zedler 2000, Boesch 2006, Gawlik 2006). We now recognize that restoration of ecosystem function requires an understanding of how other factors, such as hydrological and geomorphic regimes, soil properties, topography, disturbance regimes, landscape setting, seed banks, and invasive species, can limit the restoration process (Zedler 2000). Continued experimentation and evaluation of restoration projects, at multiple spatial and temporal scales, is needed to refine restoration techniques and to improve our predictive abilities of restoration outcomes.

REFERENCES

Allen, J.R.L. 1965. A review of the origin and characteristics of recent alluvial sediments. *Sedimentology* 5(2):89-191.

Ashley, G.H. 1910. Drainage Law of Tennessee. State of Tennessee, State Geological Survey, Nashville, *TN. Bulletin: Drainage reclamation in Tennessee.*

Barnes, B.V., D.R. Zak, S.R. Denton, S.H. Spurr. 1998. *Forest Ecology, 4th edition.* John Wiley & Sons, Inc. New York. Pages 445-446.

Barstow, C.J. 1971. Impact of channelization on wetland habitat in the Obion-Forked Deer Basin, Tennessee. Pages 362-376 in J.B. Trefethen (eds.) *Transactions of the Thirty-Sixth North American Wildlife and Natural Resources Conference.* Wildlife Management Institute, Washington, D.C.

Bedford, B. 1999. Cumulative effects on wetland landscapes: links to wetland restoration in the United States and southern Canada. *Wetlands* 19:775-788.

Begon, M., J.L. Harper, and C.R. Townsend. 1990. *Ecology: Individuals, populations and communities. Second edition*, Blackwell Scientific Publications, Boston , Mass.

Bloom, A.L. 1978. *Geomorphology: A systematic analysis of late Cenozoic landforms.* Prentice-Hall, Inc. New Jersey.

Boesch, D.F. 2006. Scientific requirements for ecosystem-based management in the restoration of Chesapeake Bay and Coastal Louisiana. *Ecological Engineering* 26:6-26.

Boulton, M. 2005. Spatio-tempral patterns of geomorphic adjustment in channeled tributary streams of the lower Hatchie River basin, West Tennessee. *Ph.D. dissertation. University of Tennessee,* Knoxville, TN.

Brandt, S.A. 2000. Classification of geomorphological effects downstream of dams. *Catena* 40:375-401.

Brierley, G.J. and C.P. Murn. 1997. European impacts on downstream sediment transfer and bank erosion in Cobargo catchment, *New South Wales, Australia. Catena* 31:119-136.

Brooks, A.P. and G.J. Brierley. 2004. Framing realistic river rehabilitation targets in light of altered sediment supply and transport relationships: lessons from East Gippsland, Australia. *Geomorphology* 58:107-123.

Church, M. 1995. Geomorphic response to river flow regulation: case studies and time-scale. *Regulated Rivers: Research and Management* 11:3-22.

Clark, J.R. and J. Benforado. 1981. *Wetlands of bottomland hardwood forests: Proceedings of a workshop on bottomland hardwood forest wetlands of the southeastern U.S.* Elsevier Science Publishing Company, New York, New York, USA.

Conner, W.H. and R.R. Sharitz. 2005. Forest communities of bottomlands. Pages 93-120, in L.H. Fredrickson, S.L. King, and R.M. Kaminski (eds*.), Ecology and Management of*

Bottomland Hardwood Systems: The State of Our Understanding. University of Missouri-Columbia. Gaylord Memorial Laboratory Special Publication No. 10. Puxico.

Dahm, C.N., K.W. Cummins, H.M. Valett, and R.L. Coleman. 1995. An ecosystem view of the restoration of the Kissimmee River. *Restoration Ecology* 3:225-238.

DeAngelis, D.L., L.J. Gross, M.A. Huston, W.F. Wolff, D.M Fleming, E.J. Comiskey, and S.M. Sylvester. 1998. Landscape modeling for Everglades ecosystem restoration. *Ecosystem* 1:64-75.

Diehl, T. 2000. *Shoals and valley plugs in the Hatchie River Watershed. 00-4279, USGS*, Nashville, TN.

Gawlik, D.E. 2006. The role of wildlife science in wetland ecosystem restoration: Lessons from the Everglades. *Ecological Engineering* 26:70-83.

Gilvear, D., and J.P. Bravard. 1996. Geomorphology of temperate rivers. Pages 69-97 in G.E. Petts and C. Amoros, eds. *Fluvial Hydrosystems*. Chapman & Hall, London.

Hall, R.B.W. and P.A. Harcombe. 1998. Flooding alters apparent position of floodplain saplings on a light gradient. *Ecology* 79:847-855.

Happ, S., G. Rittenhouse, and G. Dobson. 1940. Some principles of accelerated stream and valley sedimentation. Technical Bulletin 695, U.S. Department of Agriculture.

Hefner, J., and J. Brown. 1985. Wetland trends in the southeastern United States. *Wetlands* 4:1-11.

Heimann, D.C. and M.J. Roell. 2000. Sediment loads and accumulation in asmall riparian wetland system in Northern Missouri. *Wetlands* 20(2):219-231.

Heitmeyer, M., and L. Fredrickson. 1981. Do wetland conditions in the Mississippi Delta hardwoods influence mallard recruitmen? *Transactions of the North American Wildlife and Natural Resources Conference* 46:44-57.

Hidinger, L.L. and A.E. Morgan. 1912. Drainage problems of Wolf, Hatchie, and South Fork of Forked Deer Rivers, in West Tennessee. *The Resources of Tennessee* 2(6):231-249.

Hodges, J.D. 1997. Development and ecology of bottomland hardwood sites. *Forest Ecology and Management* 90:177-125.

Hodges, J.D. and G.L. Switzer. 1979. Some aspects of the ecology of southernbottomland hardwoods. Pages 360-365 in *North America's forests: gateway to opportunity. Proceedings of the Society of American Foresters and the Canadian Institute of Forestry.* Society of American Foresters, Washington, D.C.

Hohensinner, S., H. Habersack, M. Jungwirth, and G. Zauner. 2004. Reconstruction of the characteristics of a natural alluvial river-floodplain system and hydromorphological changes following human modifications: The Danube River (1812-1991). *River Research and Applications* 20:25-41.

Hoover, J.J, and K.J. Killgore. 1997. *Fish communities.* Pages 237-260 in M.G. Messina and W.H. Conner, eds. *Southern forested wetlands: ecology and management.* Lewis Publishers, Boca Raton, FL.

Hunt, C.B. 1967. *Physiography of the United States.* W.H Freeman and Co., San Francisco, USA.

Hunt, R.H., J.F. Walker, and D.P. Krabbenhoft. 1999. Characterizing hydrology and the importance of ground water discharge in natural and constructed wetlands. *Wetlands* 19:458-472.

Hunter, W., M. Carter, D. Pashley, and K. Barker. 1993. The Partners in Flight Species Prioritization Scheme. General Technical Report RM-227, *USDA Forest Service,* Fort Collins, CO.

Hupp, C.R. 1999. Relations among riparian vegetation, channel incision processes and forms, and large woody debris. Pages 219-245 in S.E. Darby and A. Simon (eds.), *Incised River Channels*. John Wiley and Sons, Chichester.

Hupp, C. 2000. Hydrology, geomorphology, and vegetation of Coastal Plain Rivers in the southeastern United States. *Hydrological Processess* 14:2991-3010.

Hupp, C., and D. Bazemore. 1993. Temporal and spatial patterns of wetland sedimentation, West Tennessee. *Journal of Hydrology* 141:179-196.

Hupp, C. and E. Morris. 1990. A dendrogeomorphic approach to measurement of sedimentation in a forest wetland, Black Swamp, Arkansas. *Wetlands* 10:107-124.

Hupp, C.R., E.R. Schenk, J.M. Richter, R.K. Peet, and P.A Townsend. *In press*. Bank erosion along the regulated lower Roanoke River, North Carolina. *Geological Society of America, Special Papers.*

Hupp, C.R. and A. Simon. 1986. Vegetation and bank-slope development. *Interagency Sedimentation Conference, 4th,* Las Vegas, Nevada. Proceedings: Interagency Committee on Water Data, Subcommittee on Sedimentation, Vol. 2:5-83 to 5-92.

Johnson, W. 2000. Tree recruitment and survival in rivers: influence of hydrological processes. *Hydrological Processes* 14:3051-3074.

Johnston, C.A., G.D. Bubenzer, G.B. Lee, F.W. Madison, and J.R. McHenry. 1984. Nutrient trapping by sediment deposition in a seasonally flooded lakeside wetland. *Journal of Environmental Quality* 13:283-290.

Junk, W.J., P.B. Bayley, and R.E. Sparks. 1989. The flood pulse concept in river-floodplain systems. Pages 110-127 in D.P. Dodge, editor. *Proceedings of the International Large River Symposium. Can. Spec. Publ. Fish. Aquat. Sci.* 106.

Keeland, B.D. and R.R. Sharitz. 1997. The effects of water-level fluctuations on weekly tree growth in a southeastern USA swamp. *American Journal of Botany 84* (1):131-139.

King, S. 1995. Effects of flooding regimes on two impounded bottomland hardwood stands. *Wetlands* 15:272-284.

Kleiss, B.A. 1996. Sediment retention in a bottomland hardwood wetland in Eastern Arkansas. *Wetlands* 16(3):321-333.

Knighton, D. 1998. *Fluvial forms and processes*, London, Arnold. Pp.383.

Kozlowski, T.T. 2002. Physiological-ecological impacts of flooding on riparian forest ecosystem. *Wetlands* 22(3):550-561.

Kroes, D.E. and C.R. Hupp. *In press*. Patterns of riparian sedimentation and subsidence along channelized and unchannelized reaches of the Pocomoke River, Maryland. *Journal of the American Water Resources Association.*

Leopold, L.B. and W.B. Langbein. 1966. River meanders. *Scientific American.* 214:60-70.

Leopold, L.B. and G.M. Wolman. 1957. River channel patterns: braided, meandering, and straight. *U.S. Geological Survey Professional Paper 282*-B, Washington, D.C.

Leopold, L.B., G.M. Wolman, and J.P. Miller. 1992. *Fluvial processes in geomorphology*. Dover Publications, Inc., New York, New York, USA. Pp.522.

Miller, N.A. 1990. Effects of permanent flooding on bottomland hardwoods and implications for water management in the Forked Deer River Floodplain. *Castanea 55:*106-112.

Miller, R.A., W.D. Hardeman, and D.S. Fullerton. 1966. Geologic map of Tennessee. *Tennessee Division of Geology*, TN.

Mitsch, W.J. (ed.). 2006. Wetland creation, restoration, and conservation: *The state of the science. Elsevier*, Amsterdam.

Mitsch, W.J. and J.W. Day, Jr. 2006. Restoration of wetlands in the Mississippi-Ohio-Missouri (MOM) River Basin: Experience and needed research. *Ecological Engineering* 26:55-69.

Mitsch, W.J. and J.G. Gosselink. 2007. *Wetlands: 4th edition*. John Wiley and Sons, Inc. New York, New York, USA.

Mitsch, W.J. and S.E. Jorgensen. 2004. *Ecological engineering and ecosystem restoration.* John Wiley & Sons, Hoboken, NJ.

Mitsch, W.J., X.Y. Wu, R.W. Nairn, P.E. Weihe, N.M. Wang, R. Deal, and C.E. Boucher. 1998. Creating and restoring wetlands. *Bioscience 48*:1019-1030.

Morgan, A.E. and S.H. McCrory. 1910. Preliminary report upon the drainage of the lands overflowed by the North and Middle Forks of the Forked Deer River and the Rutherford Fork of the Obion River in Gibson County, TN. State of Tennessee, *State Geological Survey*, Nashville, TN.

National Research Council. 1992. *Restoration of Aquatic Ecosystems*. National Academy Press, Washington, D.C.

Nilsson, C., C.A. Reidy, M. Dynesius, and C. Revenga. 2005. Fragmentation and flow regulation of the world's large river systems. *Science* 308:405-408.

Odum, E.P. 1969. The strategy of ecosystem development. *Science* 164: 262-270.

Oswalt, S.N. and S.L. King. 2005. Channelization and floodplain forests: impacts of accelerated sedimentation and valley plug formation on floodplain forests of the Middle Fork Forked Deer River, Tennessee, *USA. Forest Ecology and Management* 215:69-83.

Petts, G.E. and A.M. Gurnell. 2005. Dams and geomorphology: Research progress and future directions. *Geomorphology* 71:27-47.

Piazza, B.P., P.D. Banks, and M.K. LaPeyre. 2005. The potential for created oyster shell reefs as a sustainable shoreline protection strategy in Louisiana. *Restoration Ecology* 13:499-506.

Pickett, S.T.A., P.S. White. 1985. Natural disturbance and patch dynamics: an introduction. Pages 3-13 in S.T.A. Pickett and P.S. White, eds. *The ecology of natural disturbance and patch dynamics.* Academic Press, Orlando, FL.

Pierce, A.R. 2005. *Sedimentation, hydrology, and bottomland hardwood forest succession in altered and unaltered tributaries of the Hatch River, TN. Ph.D. dissertation.* University of Tennessee, Knoxville, TN.

Pierce, A.R. and S.L. King. 2007a. The effects of flooding and sedimentation on seed germination of two bottomland hardwood tree species. *Wetlands 27*:588-594.

Pierce, A.R. and S.L. King 2007b. The influence of valley plugs in channelized streams on floodplain sedimentation dynamics over the last century. *Wetlands 27*:631-643.

Pierce, A.R. and S.L. King. 2008. Spatial dynamics of overbank sedimentation in floodplain systems. *Geomorphology* 100:256-268.

Pringle, C. 2000. Threats to U.S. public lands from cumulative hydrologic alterations outside of their boundaries. *Ecological Applications* 10:971-989.

Richter, B.D., J.V. Baumgartner, J. Powell, and D.P. Braun. 1996. A method for assessing hydrologic alteration with ecosystems. *Conservation Biology* 10:1163-1174.

Risotto, S., and R. Turner. 1985. Annual fluctuation in abundance of the commercial fisheries of the Mississippi River and tributaries. *North American Journal of Fisheries Management* 5:557-574.

Robbins, C.H. and A. Simon. 1983. Man-induced channel adjustment in Tennessee streams. *USGS Water Resources Investigations Report* 82-4098.

Saucier, R.T. 1994. *Geomorphology and quaternary geologic history of the Lower Mississippi Valley Volume1.* US Army Corps of Engineers. Vicksburg, MS.

Schumm, S.A., M.D. Harvey, and C.C. Watson. 1984. Incised Channels: Morphology, dynamics, and control. *Water Resources Publications, Littleton,* Colorado.

Shankman, D. 1993. Channel migration and vegetation patterns in the Southeastern Coastal Plain. *Conservation Biology* 7(1):176-183.

Shankman, D. and T.B. Pugh. 1992. Discharge response to channelization for a Coastal Plain stream. *Wetlands* 12(3):157-162.

Sharitz, R. and W. Mitsch. 1993. Southern floodplain forests. *Biodiversity of Southeastern United States/Lowland Terrestrial Communities.* W. Martin, S. Boyce and A. Echternacht, John Wiley & Sons Inc.: 311-371.

Shelford, V.E. 1954. Some lower Mississippi valley biotic communities: their age and elevation. *Ecology* 35(2):126-142.

Simenstad, C., D. Reed, and M. Ford. 2006. When is restoration not? Incorporating landscape-scale processes to restore self-sustaining ecosystems in coastal wetland restoration. *Ecological Engineering* 26:27-39.

Simon, A. 1994. Gradation processes and channel evolution in modified West Tennessee streams: process, response, and form. *USGS professional paper 1470.* USA, Washington, D.C.

Simon, A. and C.R. Hupp. 1987. Geomorphic and vegetative recovery processes along modified Tennessee streams: an interdisciplinary approach to disturbed fluvial systems. Pages 251-262 in *Forest Hydrology and Watershed Management (Proceedings of the Vancouver Symposium, August 1987. International Association of Scientific Hydrology.*

Simon, A. and C.R. Hupp. 1992. Geomorphic and vegetative recovery processes along modified stream channels of West Tennessee. *U.S. Geological Survey, Nashville,* TN.

Simon, A. and C.H. Robbins. 1987. Man-induced gradient adjustment of the South Fork Forked Deer River, West Tennessee. *Environmental Geology and Water Sciences* 9(2):108-118.

Soil Conservation Service. 1977. Land treatment plan for erosion control and water quality improvement in the Obion-forked Deer Basin River. *USDA, SCS,* Nashville, TN.

Speer, P.R., W.J. Perry, J.A. McCabe, O.G. Lara, and others. 1965. Low-flow characteristics of streams in the Mississippi embayment in Tennessee, Kentucky, and Illinois, with a section of quality of water, by H.G. Jeffery: *USGS professional paper* 448-H.

Streng, D., J. Glitzenstein, and P. Harcombe. 1989. Woody seedling dynamics in an east Texas floodplain forest. *Ecological Monographs* 59:177-204.

Tucci, P. and G.E. Hileman. 1992. Potential effects of dredging the South Fork Obion River on ground-water levels near Sidonia, Weakley County, Tennessee. *U.S. Geological Survey, Water-Resources Investigations Report 90*-4041.

Turner, R.E., S.W. Forsythe, and N.J. Craig. 1981. Bottomland hardwood forest land resources of the southeastern U.S. Pages 13-43 in J.R. Clark and J. Benforado, eds. *Wetlands of Bottomland Hardwood Forests: Proceedings of a workshop on bottomland hardwood forest wetlands of the southeastern U.S.* Elsevier Science Publishing Company, New York, NY.

U.S. Department of Agriculture Soil Conservation Service. 1970. *Hatchie River basin survey report, Tennessee and Mississippi: U.S. Department of Agriculuture Soil Conservation Service.*

U.S. Department of Agriculture Soil Conservation Service. 1986. Sediment transport analysis report, Hatchie River Basin special study, Tennessee and Mississippi: *U.S. Department of Agriculture Soil Conservation Service,* p. 17.

van der Valk, A.G., S.D. Swanson, and R.F. Nuss. 1983. The response of plant species to burial in three types of Alaskan wetlands. *Can. J. Bot.* 61:1150-1164.

van der Valk, A.G. 1998. Sucession theory and restoration of wetland vegetation. Pages 657-667 in A.J. McComb and J. Davis (eds.), *In Wetlands for the Future.* Gleneagles Publishing, Adelaide, South Australia.

Walls, R.L., D.H. Wardrop, and R.P. Brooks. 2005. The impact of experimental sedimentation and flooding on the growth and germination of floodplain trees. *Plant Ecology* 176:203-213.

Ward, J.V. and K. Tockner. 2001. Biodiversity: towards a unifying theme for river ecology. *Freshwater Biology* 46:807-819.

Wells, A.R. 2004. Integrating geographic information systems and remote sensing with spatial econometric and mixed logit models for environmental evaluation. *Doctoral Dissertation.* University of Tennessee, Knoxville, TN.

Wharton, C.H. and M.M. Brinson. 1979. Characteristics of southeastern river systems. Pages 32-40, in R.R. Johnson and J.F. McCormick (eds.), Strategies for protection and management of floodplain wetlands and other riparian ecosystems. *U.S. Forest Service Publ.* GTR-WO-12, Washington, D.C.

Wharton, C.H., W.M. Kitchens, E.C. Pendleton, and T.W. Sipe. 1982. The ecology of bottomland hardwood swamps of the southeast: a community profile. *Publ. No. FWS/OBS-81/37, U.S. Fish and Wildlife Service,* Washington, D.C.

Wilder, T.C. 1998. A comparison of mature bottomland hardwood forests in natural and altered settings in West Tennessee. *M.S. Thesis. Tennessee Technological University,* Cookeville, TN.

Wilen, B., and W. Frayer. 1990. Status and trends of U.S. wetlands and deepwater habitats. *Forest Ecology and Management* 33/34:181-192.

Williams, G.P. and M.G. Wolman. 1984. Downstream effects of dams on alluvial rivers. *U.S. Geological Survey Professional Paper 1286,* Washington, D.C.

Wolfe, W.J. and T.H. Diehl. 1993. Recent sedimentation and surface-water flow patterns on the flood plain of the North Fork Forked Deer River, Dyer County, Tennessee. USGS, in cooperation with TWRA, Nashville, TN. *Water Resources Investigations Report 92-4082.*

Zedler, J.B. 2000. Progress in wetland restoration ecology. *Trends in Ecology and Evolution 15:402-407.*

In: Ecological Restoration
Editors: George H. Pardue and Thomas K. Olvera

ISBN 978-1-60741-013-3
© 2009 Nova Science Publishers, Inc.

Chapter 6

RESTORATION, REHABILITATION AND GARDENING IN MEDITERRANEAN STREAM ECOSYSTEMS

Jesús D. Ortiz and Gora Merseburger
Rheos ecology, Camí de Valls 81-87, E-43204 Reus, Tarragona, Catalonia, Spain
www.rheosecology.com

ABSTRACT

Humans have been exploiting stream ecosystems for millenniums. During the last two centuries, the demographic explosion resulted in the overexploitation of natural resources and channel alterations in rivers and streams from worldwide. A large number of human threats lead to ecosystem depletion. It was not until the end of the 20th century that population started to become aware of the negative consequences of their development, and thus, arose the first attempts to restore fluvial environments. Restoration in stream ecosystems entails the return to former conditions, including water chemistry, hydrologic regime, geomorphology, species composition and riparian vegetation. However, in some cases, stream ecosystems cannot be restored because it becomes too expensive and difficult or simply cannot be performed. In these cases, several rehabilitation measures could be adopted to enhance some stream ecosystem services (e.g., self-purification) and meet many stakeholder goals and objectives. Both terms, restoration and rehabilitation, can be easily confused by inexperienced water managers with the finding of esthetic values, otherwise called gardening. These erroneous restoration measures are usually restricted to merely planting trees in the riparian zone but can also involve major channel modifications. Consequently, the resulting ecosystem will need continuous intervention and the benefits of ecosystem services are neglected. To overcome this misunderstanding, restoration and rehabilitation projects have to be developed according to an appropriate knowledge of the ecological theory considering species life history, habitat template and spatio-temporal scope. All these characteristics are strongly influenced by the climate setting, which takes special relevance in the Mediterranean region because of the marked seasonality and water scarcity.

INTRODUCTION

From their outset, humans have considerably modified their environment, frequently carrying severe consequences that were in detrimental of themselves. The European Environmental Agency estimated that more than 50% of flowing freshwaters in the European Union underwent significant eutrophication by the end of the 20th century [Crouzet *et al.* 1999]. This percentage can become much higher as soon as other human disturbances are considered. In addition to eutrophication, the list of human threats to aquatic ecosystems is quite wide: weir and dam construction, canalization, water diversion and withdrawal, eutrophication (nutrients, and organic matter), pollution (garbage, biocides, drugs and chemicals), bank erosion, riparian clear-cut, introduction of exotic species, etc. Climate conditions in the Mediterranean region, characterized by soft, wet winters and hot, dry summers, make it very suitable for human settlement in comparison with other regions of the world. Water resources in the Mediterranean region are limited and, consequently, aquatic ecosystems become particularly susceptible to human impacts [Gasith & Resh 1999]. In addition, the marked seasonal and interannual variability in rainfall can also have severe consequences in stream ecosystems through floods [Lake 2000] and droughts [Lake 2003]. In addition, these natural disturbances also interact with human derived disturbances (e.g. dilution of pollutants, feasibility of fish passes, etc.). Nevertheless, considering the dramatic effects of many human threats on stream ecosystems [Paul & Meyer 2001], we are facing the challenge of continue growing without exhausting our own resources, including water, and avoiding conflicts among us. The negative consequences resulting from human development on aquatic ecosystems did not start to be evident until they become a serious problem for human health, as documented for the Thames River [Allan 1995]. Several years after, awareness become a growing trend among citizens regarding water courses not only as a resource but also as a focus of biodiversity. This change of mentality arose with the first more or less successful attempts to restore fluvial environments. However, still today most people, including citizens, but also stakeholders, managers and politicians, continue threatening water courses because are not knowledgeable about the real effects of human disturbances on the structure and function (i.e., capacity to retain and transform nutrients, etc.) of these fragile ecosystems. Furthermore, those people are sometimes more interested on the derived propaganda of the green actions, leaving the long-term success in second term. Within this context, the aim of this manuscript is to make some criticisms against misunderstandings and weakness of current restoration projects in stream ecosystems and provide some directions according to scientific knowledge to improve the development of successful restoration projects. Although we have focused this chapter on Mediterranean streams, most concepts can be applied to all ecoregions and type of ecosystems.

RESTORATION, REHABILITATION AND GARDENING

In general, the term restoration can be defined as the return to former conditions, including water chemistry, hydrologic regime, geomorphology, species composition and riparian vegetation.

When applied to stream ecosystems, this definition can become somehow ambiguous. Firstly, there could be some conflict in determining what exactly means "former conditions" for a given ecosystem, unless all human disturbances are rather recent. For some people the reference scenario could be found in the memory of the oldest citizens while a more restrictive perspective will probably raise prehistoric ages arguing that even the first big civilizations significantly altered the ecosystems. Secondly, some stream characteristics are in continuous transformation, such as geomorphology, or could not be compared to previous conditions because they were not studied before, such as species composition. And, thirdly, the extent of the ecosystem is somehow difficult to determine because stream ecosystems are tightly linked to their watersheds more than just the contiguous area [Roy *et al.* 2003]. All these inconveniences mean that the original conditions, to whatever they are referred, cannot be technically restored. The solution to this uncertainty passes through trying to improve ecosystems so that their structure and function becomes as natural as possible.

In some cases, restoration measures become excessively expensive and difficult or simply cannot be performed because original ecosystems and their dynamics are irretrievable [Bernhardt & Palmer 2007]. The heavy modifications made in most stream reaches flowing through urbanized areas represent a press disturbance that is generally difficult to restore [Allan 1995, Lake 2000]. In these cases, restoration can become somehow ambiguous. However, some rehabilitation measures could be adopted to enhance ecosystem services and life quality of citizens at the same time.

Both terms, restoration and rehabilitation, can be easily confused by inexperienced water managers with socioeconomic interests or the mere finding of esthetic values, otherwise called gardening. These erroneous restoration measures are usually restricted to just planting trees in the riparian zone and habilitate picnic areas, but can also involve major channel modifications and reintroduction, or introduction, of charismatic species. Consequently, the resulting ecosystem will need continuous intervention while most benefits of ecosystem services are neglected. The main problem of gardening projects is that after thousands of years destroying natural ecosystems, nowadays there is a fashion of awareness and the terms restoration and rehabilitation sound pretty sweet, while gardening seems much less gorgeous. With this, we are not meaning that gardening has to be avoided. On the contrary, we think that these practices have to be more common in public areas according to the potential requirements of the population for leisure and other social uses. The message here is that each project has to be called with its own name. Gardening projects can easily be camouflaged behind the terms restoration or rehabilitation. Inversely, restoration and rehabilitation can also easily fail into gardening measures. Ignorance is often the most common cause of these misunderstandings, but economic and media interests are common as well. The most contemptible case is when private owners or public administrations become able to use public funding for gardening under the name of restoration or rehabilitation to obtain indirect benefits such as customers for a rural house. The boundary between restoration and rehabilitation could become somehow diffuse in few cases, while any experienced ecologist can easily recognize gardening projects because they are focused on beauty standards instead of on ecosystem function. As said before, gardening is mainly focused on socioeconomic aspects while restoration and rehabilitation measures are directed to recover the structure and function of natural ecosystems. In this sense, special attention has to be taken to inputs of pollutants, water extraction, canalization, riparian vegetation, transversal barriers, and species introductions.

WATER QUALITY AND QUANTITY

Water quality and scarcity are by far the most common threaten to aquatic ecosystems worldwide. Agriculture and urban sewage, in this order, represent the major causes of water impairment in the European Union [Nixon *et al.* 2003]. Both activities release a number of pollutants into aquatic ecosystems including biocides, drugs, chemicals, nutrients and organic matter. All pollutants can cause severe impairments into freshwater ecosystems but the most common threats are caused by nutrients and organic matter. An excess of nutrients and organic matter can lead to eutrophication, having adverse effects on the ecology of stream ecosystems and limiting the suitability of water for human use [Paul & Meyer 2001]. Measures to mitigate the effects of agriculture on water quality of the receiving aquatic ecosystems are easy to list but difficult to apply. Restoration measures at reach scale are fated to fail because diffuse sources of nutrients are not mitigated [Moerke *et al.* 2004]. Therefore, effective restoration plans require actions at catchment scale but are hardly ever applied because of the effort and economic invest that has to be assumed [Allan 1995]. Among these measures, the most supported by land users are the development and application of latest technologies and strategies to increase the efficiency of forages and fertilizers. Other procedures, such as avoiding farming overloads, or the restoration of riparian vegetation are hardly ever considered. The efficiency of nutrient retention by riparian roots is fairly proven [Lowrance *et al.* 1984, Peterjohn & Correll 1984]. However, the subsequent drop of cultivable lands, the need of catchment scale measures [Harrison *et al.* 2004, Moerke *et al.* 2004], and the excessively long periods required to obtain perceptible results [Lake 2005, Lengsfeld & Gelbrecht 2002] dissuade private owners and administrations. Therefore, the benefits of these otherwise called ecosystem services [Constanza *et al.* 1997] are neglected [Fisher *et al.* 1998, Lowrance *et al.* 1984, Schade *et al.* 2005].

In front of the pervasive growing of population, the only ways by which the effects of wastewater inputs to aquatic ecosystems could be lessen are the awareness of population regarding water consumption and the increase of the number, capacity and efficiency of wastewaters treatment plants. The use of adequate physicochemical and/or biological tertiary treatments have demonstrated satisfactory nutrient reduction rates in wastewater that help to the preservation of natural aquatic ecosystems [e.g. Gücker *et al.* 2006]. In cases of low population densities and/or funding, the use of at least wetlands or septic systems is strongly recommended to avoid direct inputs of crude wastewater and the consequent degradation of water quality [Kivaisi 2001]. In addition to domestic sewage, industrial wastewater inputs are also susceptible to cause severe injures to aquatic ecosystems through extreme pH values, organic pollution, salts or heavy metals among other components [Hynes 1978, Rabeni *et al.* 1985, Wiederholm 1984]. In any case, once the sources of nutrient enrichment are mitigated, stream ecosystems will rapidly be recovered from their effects. In this sense, the hydrologic forces play an important role by continuously cleaning all pollutants retained in the substrate. The stream geomorphology is also an important feature to be taken into account in self-purification processes [Martí *et al.* 2006] although it may result somehow difficult to manage. Furthermore, in Mediterranean stream ecosystems flooding and drying events can accelerate those physical self-purifying processes because act as reset mechanisms [Ortiz & Puig 2007].

The scarcity and irregularity of precipitations in the Mediterranean region bring to many streams to become temporary or ephemeral. In addition, overexploitation is often translated in that water courses get dry even more often than under natural conditions and can strongly influence their natural dynamics [Gasith & Resh 1999]. After several years of restrictions, most Mediterranean countries have developed striking strategies to ensure water availability for citizens but even the most environmentalist government does forget the integrity of natural ecosystems. To diminish the effects of human induced dry outs certain organizations are currently developing techniques to discharge regenerated wastewaters into stream ecosystems. This measure is exemplified in Tossa Creek, where the use of regenerated water attenuated the effects of groundwater overexploitation and contributed to certain recovery towards the original hydrologic regime [Ortiz et al. 2008]. In this case, the indirect input of regenerated water eliminated some sensitive taxa, but the loss of richness was not significant and was compensated by the decrease of the degree of flow temporality.

In-Stream Habitat

Stream canalization involves significant alteration of the ecosystem with important loss of in-stream habitat, especially if it is made with concrete [Woodcock et al. 2007]. It is understandable that urban areas with buildings that were located so close to the stream channel have to be protected from flooding by contention walls. However, stream canalization and reduction of morphological heterogeneity (i.e. meandering, in-stream islands, blocks, fallen trees, etc.) may not suppose any advantage against inundation events while have negative effects on the structure and function of stream ecosystems. Riparian vegetation plays an important role as a refuge for adult flight stages of aquatic insects, birds and mammals among other organisms, but is also in charge of bank stabilization and sediment retention. In addition, submersed roots and fallen leaves, branches and trees provide a complex matrix where many aquatic organisms could find shelter and food resources. In most cases, fallen exotic trees become as good habitats as native trees does, and therefore leaving felt trees into the shore is a suitable fate for eliminated species in restoration projects. However, inexperienced water managers could misunderstand the relevance of fallen trees into aquatic ecosystems and cut trunks into small logs with the aim of increasing in-stream habitat. The result is that those small logs float and are transported downstream by stream flow, while few organisms are able or aim to attach to floating moving elements. Moreover, these small logs are likely to be retained in turbines or other human engines located downstream and cause inconveniences. Dams, weirs and road culverts not only alter stream connectivity (see below) but also natural processes of erosion and sedimentation and can also impoverish in-stream habitat [Orr et al. 2006]. On the other hand, garbage such as shopping trolleys, bricks or plastic bags, can offer surface to colonize for a number of aquatic organisms, including algae and invertebrates. These artificial substrata deeply break visual esthetics and release toxic compounds but, surprisingly, can become of special relevance in urban streams and canals dominated by concrete or fine sediments [Ortiz et al. 2008].

RIPARIAN VEGETATION

Riparian vegetation has been seen to affect aquatic organisms [Stanford 1998, Ward 1992] and play an important role on the stream structure and function [Cuffney *et al.* 1990, Wallace & Webster 1996]. The implications of the riparian buffer strip on stream water quality are numerous: protection from soil erosion, nutrient retention for diffuse sources of nutrients, regulator of eutrophication through canopy shadow, attenuation of floods, etc. In addition, riparian vegetation is also tightly linked to emergent insects with potentially important implications for nutrient dynamics [Grimm 1988, Sanzone *et al.* 2003, Huryn & Wallace 2000] and act as a genetic corridor for fauna and flora among other terrestrial ecosystems [Wissmar 2004].

Most restoration and rehabilitation projects are restricted to the riparian vegetation because it is much easier and visible working with trees than with other components. These kinds of projects are often focused on eliminating exotic species and planting cultivated trees and are exposed to the risk of become mere gardening projects. Of course, prior clear cut of invasive exotic species could be necessary. Nevertheless, all planted trees that survive the plantation (usually about a 70%) will be foreign varieties of native species of the same age and regularly distributed within the planting area (e.g. 3 x 3 m). If plantations are strictly necessary, they have to be performed by using seeds collected from the surroundings and planting trees of different ages according to a randomly distribution, such as copying natural ecosystems.

In most contemporary urban landscapes, natural riparian forests can represent serious inundation risk, but some trees and certain understory patches will offer shadow and esthetic values that could enhance the interaction with citizens, in addition to shelter for birds and a buffer strip to diffuse sources of nutrient pollution. Moreover, the accomplishment of good quality standards of water [Achleitner *et al.* 2005], habitat and riparian vegetation will imply the prevalence of innocuous insects (mayflies, caddisflies, stoneflies) instead of annoying big stoles of midges, biting mosquitoes or blackflies.

STREAM CONNECTIVITY

Other important threats to aquatic fauna are boundaries that disrupt longitudinal, transversal and vertical connectivity [Blakely *et al.* 2006, Kroes *et al.* 2006]. Water quality, dams, weirs gauging stations, and road culverts are the most common boundaries to longitudinal connectivity for aquatic fauna. Longitudinal connectivity is an issue of special relevance for migratory fish species (e.g. sturgeon, salmon, eel, etc) to fulfill successful life cycles. As said before, the problems driven by water quality can be ameliorated by reducing nutrient and pollutants discharges into streams. Dams and weirs built for power stations are commonly safeguarded under the label of being renewable energy resources because they do not imply direct CO_2 emissions. In Catalonia, there are more than one thousand dams and weirs and no more than fifty fish passes, most of which are useless such as stairs or long shallow ramps. The reason is that nearly all fish passes were designed without considering scientific knowledge regarding fish capabilities. The preferred option for dams and weirs is their removal when they are dispensable or could be replaced by ponds or reservoirs adjacent to the stream course. Most gauging stations are strictly necessary for the management of water resources for human use, but are often easy to replace by less intrusive and effective methods of flow measurement. The removal of a dam or a weir could entail other undesirable short-term disturbances, but the ecologic benefits at long-term scale do compensate by far the caused damages [Saunders & Kalff 2001, Stanley & Doyle 2002]. If this first choice could not be adopted, the negative effects of longitudinal boundaries can be attenuated to a certain extent by establishing fish passes [Kroes *et al.* 2006] that can also be satisfactory for invertebrates. In this case, the preferred mechanism will be the less aggressive to obtain the best effectiveness and efficiency. Hence, bypass channels and ramps better than fish ways and better than fish lifts. Road culverts, in contrast, can be easily replaced by less aggressive, and not necessarily more expensive, structures such as bridges [Blakely *et al.* 2006]. Connectivity restoration in stream ecosystems is acknowledged as having significant amelioration of stream processes and patterns. However, the elimination of certain longitudinal boundaries can also enhance the spread of exotic species and contribute to the homogenization of aquatic biota [Rahel 2007]. Certain organizations decided to promote longitudinal connectivity despise of the distribution of native and invasive species. However, certain barriers are in charge of the preservation of original stream communities, while its mitigation will certainly contribute to the spread of exotic species from downstream reaches and the decay of autochthonous populations. Considering the large number of barriers affecting our streams, the preservation of original communities has to be decisive in order to establish connectivity priorities.

Transversal connectivity refers to the existing links in terms of energy, nutrients and organisms between stream channels and the adjacent riparian zone [Sanzone *et al.* 2003, Schade *et al.* 2005]. Its relevance for aquatic biota and the challenges related to its restoration have been previously discussed in this chapter. Vertical connectivity between surface water and the hyporheic zone is much less considered in restoration plans [Jansson *et al.* 2007], probably because its relevance to aquatic ecosystems is much less recognized. However, the role of the hyporheic zone is crucial for hydrology, solute dynamics, and refuge from floods [Boulton *et al.* 1998, Marshall & Hall 2004, Olsen & Townsend 2005, Valett *et al.* 1994]. The major causes of vertical connectivity disruption are streambed canalization, and clogging by sediments or organic matter derived of agricultural activities and urban wastewaters.

EXOTIC SPECIES

The colonization of new areas by species is a natural phenomenon that has been happening countless times from the beginning of live on the earth. Humans celebrate the arrival of new species when it was achieved by natural dispersal mechanisms while regret the same achievement when it was derived from human activities. The ecological consequences of the arrival for the receiving ecosystem are likely to be similar in both cases, including species extinction, major habitat alterations, resources depletion, damages to human infrastructures, etc. The only difference lies in the dispersal mechanisms. The main problem is that human driven species introductions are much more frequent and sudden and often goes beyond the natural dispersal capacities of most species. At world scale, recurrent human derived introductions are viewed as a mechanism of genetic homogenization because few species are reaching a worldwide distribution while much other are fatten to the extinction. The list of human harmful introductions in Catalonia is quite wide: black locust (*Robinia pseudoacacia*), giant reed (*Arundo donax*), red crayfish (*Procambarus clarkii*), zebra mussel (*Dreissena polymorpha*), an unknown number of aquatic snails (e.g. genus *Physa* or *Lymnea*), catfish (*Amerulus melas*), bleak (*Alburnus alburnus*), and a long etc. There are few things to do against introduced species once they are already naturalized, especially for those species with high dispersal capacities. The complete eradication of introduced species is often unattainable because they are widespread and the extremely high economic and human effort. Conventional methods can only allow the elimination of certain introduced species from confined sites that are isolated by natural or artificial boundaries. In those cases, restoration projects have to take special care in evaluating the risk of recolonization because all investments could be squandered in few years. Therefore, the best choice is often the adoption of measures to make difficult new introductions together with the restoration and conservation of original habitats to favor the competitiveness of native species. Some researches pointed out the possibility of using specific diseases to fight against certain harmful introduced species. However, this innovative technique could represent a hazard for natural ecosystems because the effects of genetic introductions are difficult to test in complex ecosystems as Mediterranean streams are.

SCOPES FOR FUTURE PROJECTS

Most restoration and rehabilitation projects are restricted to the riparian zone because it is much easier and visible working with trees. The objectives of these projects have to become beyond just improving the riparian vegetation and necessarily entail the implementation of measures considering not only water quality but also the stream channel. Most stream restoration projects fail when considering only one or few aspects in human altered ecosystems [e.g. Harrison *et al.* 2004, Jungwirth *et al.* 2002, Moerke *et al.* 2004]. Such actions have important deficits because they were not planned according to ecological theory considering species life history, habitat template and spatio-temporal scope [Hughes 2007, Jansson *et al.* 2007, Lake *et al.* 2007]. An appropriate knowledge of the structure and function of stream ecosystems is critical to the development of successful restoration projects [Ryder & Miller 2005]. These measures are addressed to increase habitat diversity, improve longitudinal, lateral and vertical connectivity [Stanford 1998, Wissmar 2004], and enhance ecosystem services (i.e. self-purification) [Constanza *et al.* 1997]. Therefore, the measures to achieve could be as simple, or as complex, as retreating all human influences. The idea of an ecosystem devoid of any human influence in any developed country could easily derive in a utopia. Moreover, many other animals disposal their wastes and get profit from stream ecosystems. One of the most noticeable is the case of beavers, able to build dams that can dry downstream reaches by cutting large numbers of trees. Therefore, the aims of restoration projects have to be reasonable and be limited to remove or mitigate the most significant human disturbances that allow stream ecosystems be driven by natural forces. The restoration measures have to be as less intrusive as possible. Some disturbances can be mitigated without directly interfering in the stream ecosystem, while other will entail undesirable short-term disturbances or cannot be accomplished. Once all significant disturbances are mitigated, the next step in a desirable restoration project is just let nature work. This implicitly means that the actions do not require any periodic maintenance, which in turn concerns the budget of the project. Within this context, plantations and reintroductions have to be avoided as much as possible because they will certainly put up the budget and carry certain ecologic problems. As said before, the success of plantations is much lower than natural colonization and something similar occurs with reintroductions. The survival rates of reintroduced individuals born in captivity usually range about a 10%. Most of the released individuals dye because they are too young or domestic and, therefore, less competitive against their born-in-nature competitors or lack the skills to find enough food and shelter. If habitat conditions (including water quality, habitat heterogeneity and connectivity) are appropriate, the species that could life in there will arrive soon or late. This statement have been demonstrated by the recent arrival behind many years disappeared of few wolves to North Catalonia and one monk seal to the Balearic islands or the spread of the otter along many streams in Catalonia after a lot of years of regression. Therefore, reintroductions and plantations have to be restricted to those cases in which the source of colonizers is too much far or isolated or in the case of strongly endangered species. In those exceptional cases, reintroductions have to be gradual, releasing few individuals at a time and ensuring the adequacy of the habitat and the success of the reintroduced populations. This would be the case of the Iberian crayfish (*Austropotamobius pallipes*) strongly endangered because of the lethal effects of the introduced fungi *Aphanomyces astaci*. From a selfish point of view, we all, and specially stakeholders, want to see the forest but it is much more ecologically sound to respect the dynamics of natural ecosystems. This means that some restorations will probably take longer, but at geologic scale, the delay will not be significantly different. The most important is that the damage caused by human disturbances does not fall into irreversibility, such as species extinction. Stream ecosystems, and their surroundings, are known to present a high resilience that allows them to recover from almost any disturbance. These self-restoration processes are driven by succession processes in the way that every community is progressively followed by another that is more evenly structured. This way, recolonizations are more gradual and achieved according to the availability of habitat and food resources. However, in contrast to streams of other ecoregions, Mediterranean stream ecosystems rarely achieve a steady state for a long period because of the marked seasonality and the frequency of flooding and drying. These natural disturbances continuously disrupt stream communities and act as a reset mechanism for succession processes [Townsend *et al.* 1997].

Although the aims of rehabilitation projects will intrinsically be more modest than restoration ones, the principles and scopes have to be common. It is understandable that urban areas with buildings that were located so close to the stream channel have to be protected from flooding by contention walls. However, some measures such as complete elimination of riparian vegetation, stream canalization and reduction of morphological heterogeneity (i.e. meandering, in-stream islands, blocks, fallen trees, etc.) may not suppose any advantage against inundation events. Furthermore, these elements can offer shelter to aquatic organisms [Harrison *et al.* 2004, Woodcock *et al.* 2007], self-purification [Niezgoda & Johnson 2005], and esthetic values and leisure to citizens [Woolsey *et al.* 2007]. Indeed, these near-to-natural ecosystems will require certain maintenance measures. This philosophy of merging buildings with nature is the basis of the conceptual framework of ecocities [Register 1987]. It is expected that this new conception of urban landscapes will start to be implemented in few decades. However the success will depend on the application of current knowledge about natural ecosystems without falling into the development of garden cities with many plants and birds but still with serious ecological problems and wasting the benefits of ecosystem services. On the other hand, independently of how green could be cities in the future, the world population is facing a markedly growth that is expected to continue at least for some decades further. This excessive growth will irremediably increase demand of natural resources supply, including water, energy and raw materials, and production of pollutants. Therefore, if the forecast is not mistaken, streams and all other natural ecosystems will be seriously stressed. Nowadays we are facing the challenge of continue growing without exhausting our own resources. However, human history and a number of studies about animal populations demonstrate us that every remedy is destined to be unsuccessful because our planet is not able to support our constant press. Consequently, the most suitable solution at long-term is the reduction of the world population before collapse by having less children and having them at higher ages (ideally between 30 and 35 years). This solution is often considered a taboo, even among environmentalists. Obviously, this measure will carry undesirable consequences such a slowing down of the economy, the development or the research. All benefits and inconveniences have to be balanced.

Finally, a successful restoration project as a type of management strategy has to include the following eight steps: 1) Identification of the problematic. 2) Assessment of the ecosystem status. 3) Setting of the goals and objectives. 4) Decision making. 5) Application of management measures. 6) Monitoring of the environmental and social repercussions. 7) Evaluation of the data derived from the monitoring. 8) Dissemination of the results [Ortiz & Merseburger 2008]. Here we would like to highlight the importance of develop projects according to the adaptive management approach, that lies in a stepwise experimental approach with feedbacks that constantly improve the objectives of the restoration. Another relevant issue that has to be included in any restoration project is the monitoring, data analysis and dissemination of the success considering both ecological attributes and socio-economic aspects. This way, new findings and experiences acquired in restoration projects will not be limited to the directly involved organizations and the development of new projects will benefit from empirical data from previous experiences.

REFERENCES

Achleitner, S., Toffol, S., Engelhard, C., & Rauch, W. (2005). The European Water Framework Directive: water quality classification and implications to engineering planning. *Environmental Management*, 35, 517-525.

Allan, J. D. (1995). *Stream ecology: structure and function of running waters*. London, UK: Chapman & Hall.

Bernhardt, E. S. & Palmer, M. A. (2007). Restoring streams in an urbanizing world. *Freshwater Biology*, 52, 738-751.

Blakely, T. J., Harding, J. S., McIntosh, A. R. & Winterbourn, M. J. (2006). Barriers to the recovery of aquatic insect communities in urban streams. *Freshwater Biology*, 51, 1634-1645.

Boulton, A. J., Findlay, S., Marmonier, P., Stanley, E. H., & Valett, H. M. (1998). The functional significance of the hyporheic zone in streams and rivers. *Annual Reviews of Ecology and Systematics*, *29*, 59-81.

Constanza, R., d'Arge, R., de Groot, R., Farber, S., Grasso, M., Hannon, B., Limburg, K., Naeem, S., O'Neill, R.V., Paruelo, J., Raskin, R.G., Sutton, P., & Van den Belt, M. (1997). The value of the world's ecosystem services and natural capital. *Nature*, 387, 253–260.

Crouzet, P., Leonard, J., Nixon, S., Rees, Y., Parr, W., Laffon, L., Bogestrand, J., Kristensen, P., Lallana, C., Izzo, G., Bokn, T., Bak, J., & Lack, T. J. (1999). Nutrients in European ecosystems. *4*, 1-155.

Cuffney, T. F., Wallace, J. B., & Lugthart, G. J. (1990). Experimental evidence quantifying the role of benthic invertebrates in organic matter dynamics of headwater streams. *Freshwater Biology*, *23*, 281-299.

Fisher, S. G., Grimm, N. B., Martí, E., Holmes, R. M., & Jones, J. B. Jr. (1998). Material spiraling in stream corridors: A telescoping ecosystem model. *Ecosystems*, *1*, 19-34.

Gasith, A. & Resh, V. H. (1999). Streams in Mediterranean climate regions: abiotic influences and biotic responses to predictable seasonal events. *Annual Reviews of Ecology and Systematics*, *30*, 51-81.

Grimm, N. B. (1988). Role of macroinvertebrates in nitrogen dynamics of a desert stream. *Ecology*, *69*, 1884-1893.

Gücker, B., Brauns, M., & Push, M. T. (2006). Effects of wastewater treatment plant discharge on ecosystem structure and function of lowland streams. *Journal of the North American Benthological Society*, *25*, 313-329.

Harrison, S. S. C., Pretty, J. L., Shepherd, D., Hildrew, A. G., Smith, C., & Hey, R. D. (2004). The effect of instream rehabilitation structures on macroinvertebrates in lowland rivers. *Journal of Applied Ecology*, *41*, 1140-1154.

Hughes, J. M. (2007). Constraints on recovery: using molecular methods to study connectivity of aquatic biota in rivers and streams. *Freshwater Biology*, 52, 616-631.

Huryn, A. D. & Wallace, J. B. (2000). Life history and production of stream insects. *Annual Review of Entomology*, *45*, 83-110.

Hynes, H. B. N. (1978). Biological effects of organic matter. *The biology of polluted waters* (pp. 92-121). Cambridge, Great Britain: Liverpool University Press.

Jansson, R., Nilsson, C. & Malmqvist, B. (2007). Restoring freshwater ecosystems in riverine landscapes: the roles of connectivity and recovery processes. *Freshwater Biology*, 52, 589-596.

Jungwirth, M., Muhar, S., & Schmutz, S. (2002). Re-establishing and assessing ecological integrity in riverine landscapes. *Freshwater Biology*, *47*, 867-887.

Kivaisi, A. K. (2001). The potential for constructed wetlands for wastewater treatment and reuse in developing countries: a review. *Ecological Engineering*, *16*, 545-560.

Kroes, M. J., Gough, P., Schollema, P. P., & Wanningen, H. (2006). *From sea to source; Practical guidance for restoration of fish migration in European rivers*. Utrecht, The Netherlands: Plantijn Casparie Nieuwegein.

Lake, P. S. (2000). Disturbance, patchiness, and diversity in streams. *Journal of the North American Benthological Society*, *19*, 573-592.

Lake, P. S. (2003). Ecological effects of perturbation by drought in flowing waters. *Freshwater Biology*, *48*, 1161-1172.

Lake, P. S. (2005). Perturbation, restoration and seeking ecological sustainability in Australian flowing waters. *Hydrobiologia, 552,* 109-120.

Lake, P. S., Bond, N. & Reich, P. (2007). Linking ecological theory with stream restoration. *Freshwater Biology,* 52, 597-615.

Lengsfeld, H. & Gelbrecht, J. (2002). Restoration measures to reduce diffuse pollution in a small catchment area in NE Germany - a first evaluation.

Lowrance, R., Todd, R., Fail, J. Jr., Hendrickson, O. Jr., Leonard, R., & Asmussen, L. (1984). Riparian forests as nutrient filters in agricultural watersheds. *Bioscience, 34,* 374-377.

Marshall, M. C. & Hall, R. O. Jr. (2004). Hyporheic invertebrates affect N cycling and respiration in stream sediment microcosms. *Journal of the North American Benthological Society, 23,* 416-428.

Martí, E., Sabater, F., Riera, J. L., Merseburger, G. C., von Schiller, D., Argerich, A., Caille, F., & Fonollà, P. (2006). Fluvial nutrient dynamics in a humanized landscape. Insights from a hierarchical perspective. *Limnetica, 25,* 513-526.

Moerke, A. H., Gerard, K. J., Latimore, J. A., Hellenthal, R. A., & Lamberti, G. A. (2004). Restoration of an Indiana, USA, stream: bridging the gap between basic and applied lotic ecology. *Journal of the North American Benthological Society, 23,* 647-660.

Niezgoda, S. L. & Johnson, P. A. (2005). Improving the urban stream restoration effort: identifying critical form and processes relationships. *Environmental Management, 35,* 579-592.

Nixon, S., Trent, Z., Marcuello, C., & Lallana, C. (2003). Europe's water: An indicator-based assessment. *1/2003,* 1-98.

Olsen, D. A. & Townsend, C. R. (2005). Flood effects on invertebrates, sediments and particulate organic matter in the hyporheic zone of a gravel-bed stream. *Freshwater Biology, 50,* 839-853.

Orr, C. H., Rogers, K. L., & Stanley, E. H. (2006). Channel morphology and P uptake following removal of a small dam. *Journal of the North American Benthological Society, 25,* 556-568.

Ortiz, J. D. & Merseburger, G. (2008). Stream management in Mediterranean ecosystems. In M. S. Alonso & I. M. Rubio (Eds.), *Ecological management: new research* New York: Nova Science Publishers, Inc.

Ortiz, J. D., Merseburger, G., Martí, E., Ordeix, M., & Sabater, F. (2008). Effects of urbanization on aquatic macroinvertebrates in Mediterranean streams. In L. N. Wagner (Ed.), *Urbanization: 21st century issues and challenges* (pp. 91-132). New York: Nova Science Publishers, Inc.

Ortiz, J. D. & Puig, M. A. (2007). Point source effects on density, biomass and diversity of benthic macroinvertebrates in a Mediterranean stream. *River Research and Applications, 23,* 155-170.

Paul, M. J. & Meyer, J. L. (2001). Streams in the urban landscape. *Annual Reviews of Ecology and Systematics, 32,* 333-365.

Peterjohn, W. T. & Correll, D. L. (1984). Nutrient dynamics in an agricultural watershed: observations on the role of a riparian forest. *Ecology, 65,* 1466-1475.

Rabeni, C. F., Davies, S. P., & Gibbs, K. E. (1985). Benthic invertebrate response to pollution abatement: structural changes and functional implications. *Water Resources Bulletin, 21,* 489-497.

Rahel, F. J. (2007). Biogeographic barriers, connectivity and homogenization of freshwater faunas: it's a small world after all. *Freshwater Biology*, 52, 696-710.

Register, R. (1987). *Ecocity Berkeley: building cities for a healthy future*, North Atlantic Books

Roy, A. H., Rosemond, A. D., Paul, M. J., Leigh, D. S., & Wallace, J. B. (2003). Stream macroinvertebrate response to catchment urbanisation (Georgia, U.S.A.). *Freshwater Biology*, 48, 329-346.

Ryder, D. S. & Miller, W. (2005). Setting goals and measuring success: linking patterns and processes in stream restoration. *Hydrobiologia*, 552, 147-158.

Saunders, D. L. & Kalff J. (2001). Nitrogen retention in wetlands, lakes and rivers. *Hydrobiologia*, 443, 205-212.

Sanzone, D. M., Meyer, J. L., Martí, E., Gardiner, E. P., Tank, J., & Grimm, N. B. (2003). Carbon and nitrogen transfer from a desert stream to riparian predators. *Oecologia, 134*, 238-250.

Schade, J. D., Welter, J. R., Martí, E., & Grimm, N. B. (2005). Hydrologic exchange and N uptake by riparian vegetation in an arid-land stream. *Journal of the North American Benthological Society*, 24, 19-28.

Stanford, J. A. (1998). Rivers in the landscape: introduction to the special issue on riparian and groundwater ecology. *Freshwater Biology*, 40, 402-406.

Stanley, E. H., & Doyle M. W. (2002). A geomorphic perspective on nutrient retention following dam removal. *BioScience* 52, 693-701.

Townsend, C. R., Scarsbrook, M. R., & Dolédec, S. (1997). The intermediate disturbance hypothesis, refugia, and biodiversity in streams. *Limnology and Oceanography*, 42, 938-949.

Valett, H. M., Fisher, S. G., Grimm, N. B., & Camil, P. (1994). Vertical hydrologic exchange and ecological stability of a desert stream ecosystem. *Ecology*, 75, 548-560.

Wallace, J. B. & Webster, J. R. (1996). The role of macroinvertebrates in stream ecosystem function. *Annual Review of Entomology*, 41, 115-139.

Ward, J. V. (1992). *Aquatic insect ecology: 1. Biology and habitat*. New York: John Wiley & sons, Inc.

Wiederholm, T. (1984). Responses of aquatic insects to environmental pollution. In V. H. Resh & D. M. Rosenberg (Eds.), *The ecology of aquatic insects* (pp. 508-557). New York: Praeger.

Wissmar, R. C. (2004). Riparian corridors of Eastern Oregon and Washington: Functions and sustainability along lowland-arid to mountain gradients. *Aquatic Sciences*, 66, 373-387.

Woodcock, T. S. & Huryn, A. D. (2007). The response of macroinvertebrate production to a pollution gradient in a headwater stream. *Freshwater Biology*, 52, 177-196.

Woolsey, S., Capelli, F., Gonser, T., Hoehn, E., Hostmann, M., Junker, B., Paetzold, A., Roulier, C., Schweizer, S., Tiegs, S. D., Tockner, K., Weber, C., & Peter, A. (2007). A strategy to assess river restoration success. *Freshwater Biology*, 52, 752-769.

In: Ecological Restoration
Editors: George H. Pardue and Thomas K. Olvera

ISBN 978-1-60741-013-3
© 2009 Nova Science Publishers, Inc.

Chapter 7

Lessons that Nature Urges People to Learn: Experience from Open Drain Research

Romanas Lamsodis[1,1], Arvydas Povilaitis[2], Vaclovas Poškus[1]
and Stasys Ragauskas[1]

[1]Water Management Institute of Lithuanian University of Agriculture,
Vilainiai, Kėdainiai, Lithuania
[2]Lithuanian University of Agriculture, Kaunas, Lithuania

Abstract

Landscape changes turned drastic in 20th Century. In the agricultural, it was challenged by the need of rapid growth of agricultural production that in turn required too wet soils to be drained. Thus land reclamation significantly changed the structure of the former landscapes. The mosaic and patchiness, comprising of natural grasslands, bushes and groves, wetlands and water-bodies, as well as numerous small land-tenures, disappeared; heterogeneity was lost. The canalization of streams, which deprived them of diversity of biotic and abiotic conditions, was followed by physical destruction of riparian zones. Erosion and deflation increased; the competence of the landscape to preserve the genetic heritage of biodiversity and prevent fluvial systems from pollution as well as nutrients to be transferred into other geo-systems was restricted. Referring to the data collected over last 15 years, the attempt was made to highlight the landscape efforts to recover. It was found that these efforts involve various components of geodynamic, eolian, hydrological, biological processes leading to soil conservation, bed process stabilization, biota diversity, sedimentation prevention and water quality improvement. These alterations are also discussed as the bi-directional effect of self-renaturalisation processes in landscape on the environment and practice, and as nature's invitation to people to collaborate improving ecological situation in agro-landscapes.

---

[1] Correspondence to: e-mail* lammor@delfi.lt

INTRODUCTION

Landscape, like every other living organization, can be damaged, ill or even die. There are many patterns of dramatic man-made alterations to natural landscapes all over the Earth, which have become textbook examples of how catastrophic their consequences can be both to the environment and to society. Although all alterations had small beginnings natural processes and functions of landscapes were seriously damaged later on.

Across the centuries, structural alterations in landscape have been incessant but slow (Table 1). Marked acceleration occurred in the 20th Century and, as distinguished for Central Europe by Bernhardt *et al.* (1985; cited by Bastian *et al.*, 1993), was coupled with both the industrial and, particularly in the second half of the Century, with scientific and technological revolutions in all fields of human activity. In agriculture, landscapes changed as a result of a rapid growth in the need for farm produce, which, in turn, required soils that were too wet to be drained. Land reclamation was fundamental in significantly changing the former structure of landscapes, when assessing in local (landscape) scale (Povilaitis, 2000, 2001; Hietala-Koivu *et al.*, 2004). The traditional mosaic and patchiness of the landscape, comprising natural grasslands, bushes and groves, wetlands and water-bodies, as well as numerous small land-tenures, disappeared; the heterogeneity responsible for landscape resistance (Forman, 1995a) was lost (Figure 1). The canalisation of streams, which deprived them of the diversity of both biotic and abiotic conditions, was followed by physical destruction of the natural riparian zones which: (1) act as buffers for sediment and nutrient fluxes (Schlosser *et al.*, 1981; Rachinskas, 1983; Décamps *et al.,* 2004; Schoonover *et al.*, 2006); (2) as corridors, are indispensable for the sustainable functioning of any landscape (Forman, 1995b). Fields were greatly expanded, becoming practical for farming but at a cost. Erosion and deflation increased (Paulyukyavichyus *et al.*, 1987; Rachinskas *et al.*, 1987). The competence of the landscape, to preserve biodiversity (Naiman *et al.,* 1993) and prevent both fluvial systems from pollution and nutrients to be transferred into other geo-systems, was restricted.

Fortunately, although landscape can be damaged, it is also capable of recovery, *i.e.* rehabilitating itself. Numerous commonplace patterns demonstrate this capability involving, in various aspects, geodynamic, eolian, hydrological, biological and geochemical processes leading to soil conservation, bed processes stabilization, biota diversity, sedimentation prevention, water quality improvement and *etc.* Manifestations of the rehabilitation of landscapes can, however, negatively affect man-created structures or parts of the landscape transformed for utilization. The parameters of these structures and transformations are usually settled by calculations or technological, hydrological and agricultural requirements, therefore they need to be maintained. Right for this reason, one can consider that the conflict between man and environment is perpetuated.

This article, referring to the data collected over last 15 years, is aimed at (1) exemplifying the incessant efforts of the environment to recover its original processes and functions; (2) highlighting and discussing these efforts as nature's suggestions and invitations for people to collaborate on improving the ecological situation in agro-landscapes.

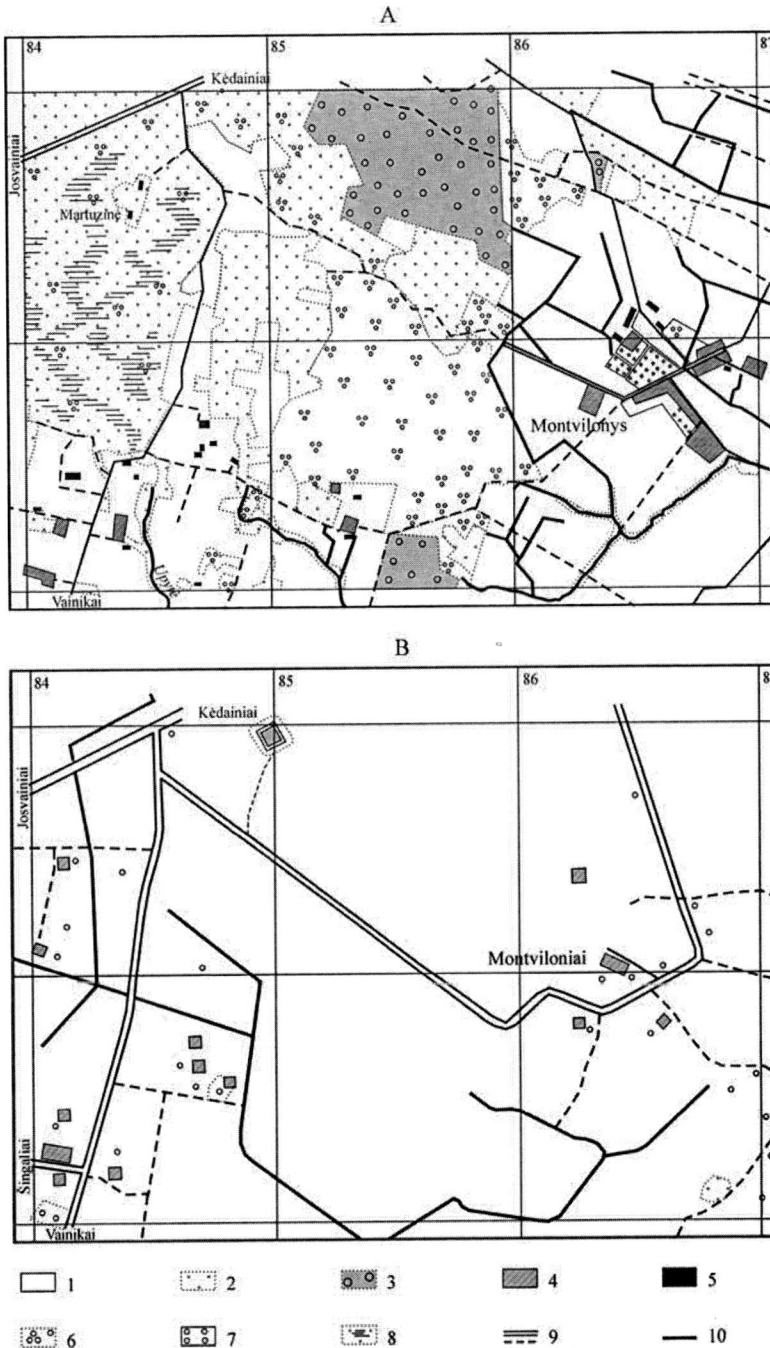

Figure 1. Two fragments of maps of 1933 (A) and 1985 (B) demonstrating the patchiness of landscape-scale alterations occurring in the same plot of the Nevėžis Plain (Middle Lithuania Lowland). The length of each side of the square coordinates equals one kilometer. Some indexes of landscape diversity were computed ('A' versus 'B'): number of patches, 23.8 *vs.* 7.8 patches per sq. km; heterogeneity after Shannon-Wiener (Palang *et al.*, 1998; Müller *et al.*, 2004), 0.83 *vs.* 0.65; evenness after Pielou (Pielou, 1975; Falk *et al.*, 2008), 0.83 *vs.* 0.93. Legend explanations: (1) arable land, (2) grassland, (3) forest, grove, (4) homestead, (5) building, (6) scrubland, (7) garden, (8) wet grassland, (9) roads, (10) open drain, intact stream. Location of the plot in territory is shown in Figure 2.

Table 1. Structural alterations of landscapes in Lithuania
over the last Millennium (regional scale)

Year/century	Landed property, %			References
	Forests	Fields	Others	
1000	56	20	24	Matulionis, 1930
1300	51	30	19	Matulionis, 1930
16th c	36–66	30–50	4–36	Eitmanavichiene,1984
1600	44	40	16	Matulionis, 1930
1900	24	66	10	Matulionis, 1930
2000	30	54	16	Agriculture…, 2000

MATERIALS, SITE DESCRIPTION AND METHODS

The data presented in the article concern the phenomena taking place within and round the open drains (hereafter drain(s)) – newly dug but mostly canalized (straightened, deepened and trapezium-shaped) natural streams of the 1st to 3rd order, and upper reaches of rivers of the higher order, for water to collect and discharge from subsurface drain systems, as well as surface runoff from rural territories – that even now remain as some of the most dynamic and sensitive parts of the agro-landscape. The data were collected during investigations carried out in various years during the period 1992–2007, to be used for diverse purposes aimed at assessing: 1) deformations of drains; 2) sediment sources and accumulation; 3) succession of vegetation cover and promise of afforestation of drain slopes; 4) beaver (*Castor fiber* L.) expansion into drains and the potential to manage the consequences; 5) water quality in forested and beaver-obstructed drains.

The investigations were conducted within a territory of about 2600 km^2, entering the Nevėžis River basin in the central part of the Middle Lithuania Lowland (Figure 2). (Lithuania is located at the place where boreal and temperate belts change together. The perennial averages here for radiation balance, precipitation and evaporation approximate respectively to 1,500MJ m^{-2}, 738mm and 512mm (Kaušyla, 1981; Lasinskas, 1981). There are four types of landscapes entering the Nevėžis River basin: clayey plains (89.4 % of total area), sandy plains (6.4 %), river valleys (3.2 %) and hilly moraine eminences (1.0 %) (Basalykas, 1977).) The territory is considered as an area of intensive farming with a high percentage of drained land (70.3 hectares per km^2), a low woodedness (23.8 hectares per km^2) and a dense net of drains (1.21 km km^{-2}).

In the territory, 545 random drain stretches were explored. The age of the drains varied from 10 to 50 years; catchment areas of the stretches ranged between 0.1–40.8 km^2; the designed dimensions of the drains (the bed width, its longitudinal gradient and slope coefficients) ranged as follows: 0.4–1.0 m, 0.5–7.5 ‰ and 1:1.5–1:2.5 respectively; the soils dominating in the tracks of the drains and in their close vicinity were checked to be of different textures, *i.e.* from sand to heavy sandy-loam, with some occurrence of peat soil. The random selection of stretches, as well as the variety of morphological conditions, allowed the study results to generalize for all the drains.

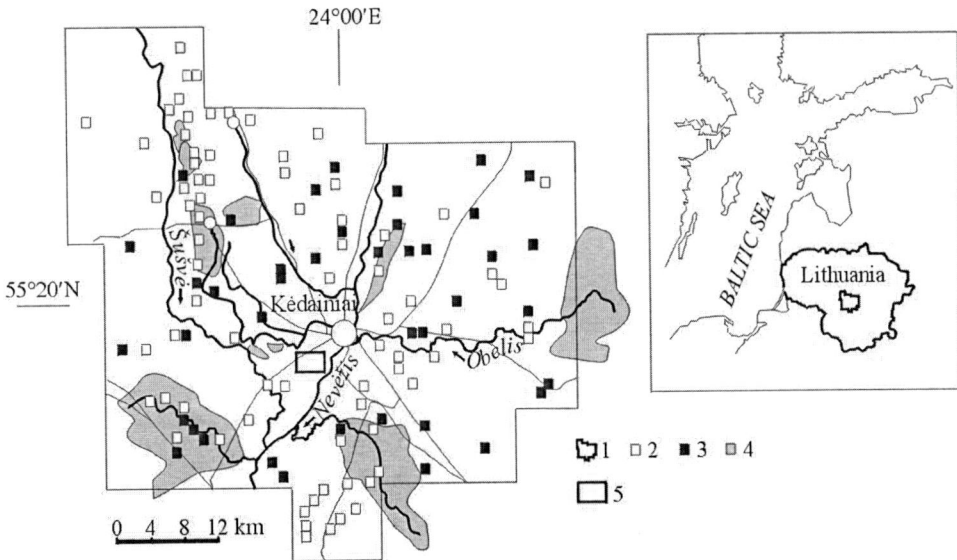

Figure 2. Location of territory (1) and randomly selected 1-km^2 squares, in each of which were five randomly chosen drain stretches for investigation of geo-morphological changes (2) and woody vegetation spread (3). Grey-toned sectors (4) mark those portions of river basins where the consequences of beaver expansion into drains were studied. (5) The plot that is analyzing in Figure 1.

All metric dimensions were taken directly. The vegetation cover was assessed by determining all species, their frequency of occurrence and density of spread.

An additional 23 drains were selected to study their water quality. The selection criteria were that drains would stretch through fields, the slopes of the upper reaches of the drains would exhibit no woody vegetation but would get overgrown downstream, or the downstream reaches would stretch into a forest where slopes usually exhibit woody vegetation. Water quality was also studied in 18 beaver-inhabited drains. Water samples were taken in the non-overgrown and overgrown stretches, and, in the case of beaver impoundments, in the drain upstream from the beaver-site, in beaver-site and in the drain downstream from the beaver-site. Water quality was assessed according to its biochemical oxygen demand (BOD) and the concentrations of dissolved oxygen (DO), total nitrogen (TN), dissolved inorganic nitrogen (DIN $\equiv$ NH$_4$-N + NO$_3$-N), total phosphorus (TP) and PO$_4$ phosphorus (PO$_4$-P). Concentrations were determined using the following methods: electro-chemical, gas diffusion, cadmium reduction, and molybdenum with stannous chloride using an "FIA Star 5012" analyzer. 200 samples were analysed from drains affected by woody vegetation and 324 from drains inhabited by beavers.

EXAMPLES OF LANDSCAPE ENVIRONMENTAL RECOVERY

There are many different examples of efforts of landscapes to restore their original processes and functions that were disturbed by human activity. It will be observed that the examples provided below are not rare; they are widespread across all drains.

Example 1: Cross-Sections of Drains

Drains are excavated significantly deeper than previous to canalisation the natural streams of comparable capacity were. However, the widths of drain beds are usually chosen to be as narrow as technologically possible, in respect of implementation and maintenance, to ensure sufficiently high flow velocities and to prevent settling of sediment during low water periods. Comparing drains to intact streams, the ratio between the bed width and depth has changed to the disadvantage of the drains; the ratio got lower resulting in the drain channel becoming more hydraulically exposed. The deeper drains both intersect more different ground strata and have more emerging seepage of groundwater onto the slope face. This results in favourable conditions for more frequent earth slumps to occur on the slope than happen with intact streams (Figure 3). Moreover, the high slopes, along with the undercutting of toes by the scouring action of flowing water, are responsible for numerous slope landslides (Figure 4).

Despite drains being continuously maintained to keep their dimensions and other parameters as designed, the dimensions and shapes of cross-sections have changed with time: the beds have widened, slopes of lower parts have got flatter whilst upper slopes have become steeper (Table 2). The cross-sections of drains are gradually being formed into roughly parabolic shapes instead of the trapezium ones.

Figure 3. Slump occurred on the slope of the Šerkšnys drain. Depth of the drain approximates to 2.75 m, designed bed width and coefficient of slopes are respectively 0.8 m and 1:2.0. (Photo: V. Poškus)

Figure 4. Landslide of the slope of the Žiedupis drain occurred following undercut of the toe by flowing water. It should be emphasized that undercuts often happen as a result of damage to the toes of slopes during the dredging of silt (Ragauskas, 2002). (Photo: V. Poškus)

Table 2. Designed and measured cross-section dimensions after prolonged operation of drains (up to 50 years)

Cross-section dimension	Part of slope	Designed value (X_o)	Measured value		Comparison ($\bar{X} - X_o$), %
			n	$\bar{X} \pm SD$	
Bed width, m	–	0.40	193	1.1±0.4	+180
Bed width, m	–	0.60	88	1.3±0.4	+112
Bed width, m	–	0.80	9	1.6±0.6	+101
Slope coefficient	Upper	1:1.5	223	1.6±0.4	+9
Slope coefficient	Lower	1:1.5	223	2.1±0.6	+27
Slope coefficient	Upper	1:2.0	250	1.7±0.4	−14
Slope coefficient	Lower	1:2.0	250	2.5±0.8	+26

Notation: n = number of measurements; $\bar{X}$ = mean; SD = standard deviation.

Example 2: Vegetation in Drains

Newly canalised streams and newly dug drains exhibit no vegetation. The initial vegetation, exclusively herbs, was intentionally introduced into drains when drain slopes were being strengthened, either by sowing a few gramineous and leguminous species or by covering with turf. In both cases, however, selected species were not adequately specialized to grow under the changing soil and moisture conditions of the down-slope. This resulted in some intentionally introduced species self-relocating, according to growing conditions, whilst those that could not specialize died, and new ones emerged (Figure 5). Over time, herbs started growing on the drain bed as well (Figure 6). In general, quite a number of various species can be found in drains. Berankienė (1997) and Lamsodis et al. (2006), indicate that herbs belonging to 39 genera and 23 families have been found growing in drains.

Figure 5. In drains, quite a number of various specialized species which thrive on poor dug or fertile, arid or excessively wet slope soils have dispersed, succeeding grassland herbs. The Graisupis drain case. (Photo: V. Poškus)

Figure 6. Once herbs begun to grow on wide bed, the drain made (over time) flow meanders. An example from the Mėkla drain. (Photo: V. Poškus)

Thirty-six woody vegetation species, belonging to 22 genera and 13 families, were found growing on drain slopes. Slopes were more overgrown if drains were closer to a forest (Figure 7). When the catchment area of the drain did not exceed 2.7 (on average 1.2; $2SD = 1.5$) km^2, some species of willow (*Salix* spp.) also chanced growing on the drain beds.

It is known that trees and shrubs suppress the growth of herbs, including helophytes. When the crowns of trees growing on the slopes of a drain have closed together to form a canopy above the bed, the helophytes no longer grow at all on the bed (Figure 8).

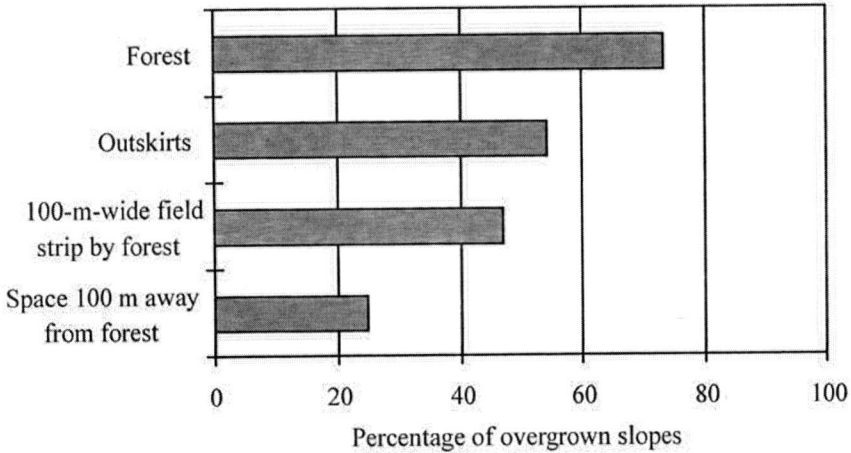

Figure 7. Percentage area of overgrown slopes of drains located at different distances from a forest. The number of species decreased with increased distance of drains from a forest; the further a drain was from a forest, deciduous trees disappeared from its slopes in the following sequence: cork elm (*Ulmus suberosa* Moench) at 200m, black alder (*Alnus glutinosa* L.), European spindle-tree (*Eonymus europaea* L.) and European bird cherry (*Padus avium* Mill.) at 400m, European ash (*Fraxinus excelsior* L.) and pedunculate oak (*Quercus robur* L.) at 500m, and gray alder (*Alnus incana* (L.) Moench) at 1000m. In field drains, no conifers were found growing at all.

Figure 8. The Graisupis-1 drain, whose slopes were intentionally planted with trees suppressing the growth of helophytes on the bed. (Photo: V. Poškus)

Example 3: Erosion and Deflation

Figure 9. Deepening of canalised streams lowered the erosion basis – the horizontal surface elevation at which the processes of erosion eventually cease. The bank of the Krimslė drain that was unprotected from erosion (no buffer strip is arranged) had a gully cut into it by concentrated surface water flowing in from an adjacent field. (Photo: V. Poškus).

Figure 10. Deflation occurred in spring in a field of grain crops next to the Paparčia drain, which has no buffer strip. It is apparent that deflation happens successfully in winter when the field is tilled and there is drifting snow without complete coverage by snow-pack. Dilys *et al.* (1997) indicated that the winter-time deflation equalled the spring-time one both in measures and consequences. (Photo: V. Poškus).

During the last 300 years in Lithuania, lands susceptible to erosion that cover more than 4600 km^2 (making up more than 7 % of the total area of the country) have lost 33 cm from the depth of the soil top-layer (Vaičys *et al.*, 1998). We have no analogous quantitative assessment of the lands affected by deflation; however, it is evident that land reclamation encouraged the processes, both by extending open areas and arable plots and by reducing the variety and number of patches of different land use. Along with this, land reclamation can be said to have given the erosion and deflation processes a push from the uplands, substantially moving them within the plains that were the most suitable places for farming in the landscape. Either process affects the A_p horizon of a soil, depriving it of the finest soil particles, nutrients and organic matter. (Referring to Račinskas *et al.*, (1986), there were 2–5 $\times 10^3$ kg of the mould, 150–300 kg of nitrogen and 10–20 kg of both phosphorus and potassium blown off per one hectare of the drained field during only one deflation event that occurred in spring of 1984 in West Lithuania.) The soils of such fields gradually become poorer, their texture becomes coarser.

Example 4: Beaver in Drains

To inhabit drains, beavers (*Castor* spp.) have to build dams and thus arrange their impoundment (Figure 11).

Figure 11. Beaver dammed up the Smilgaitis-8 drain. The height of the dam approximates to two metres. The green fence growing by the right side of the drain provided the 'building materials' for the dam. (Photo: V. Poškus)

This substantially changes a drain's hydrology, as well as its geomorphology; the ecological changes affect both the drains and their environs. In-depth investigations over the past two years have revealed that the mean values of the main dimensions of one beaver dam and one pond – dam volume, pond-water surface area and volume – averaged out at 11.3 m^3, 0.16 $\times 10^4$ m^2 and 0.88 $\times 10^3$ m^3 respectively (Lamsodis, 2008a; 2008b). *Ibidem*, it was calculated, for example, that in Lithuania (1) the part of runoff retained in beaver ponds could approximate to 0.53×10^3 m^3 in every one sq. kilometre of territory (not including losses from evaporation and infiltration); (2) geodynamic processes caused by beaver activities could induce the movement of no less than 26 m^3 of earth and other substances per every one kilometre of drains (including only the substances piled up in their dams, and the earth both washed from slopes when water was flowing round the dams (Figure 12) and borrowed from beaver holes).

Figure 12. The scour washed in drain slope when water was flowing round the beaver dam. The Aluona drain case. (Photo: V. Poškus)

Example 5: Sediment Accumulation

For the accumulation process to occur, there are two basic requirements: (1) the water flow is to be loaded with suspended sediment and (2) the flow velocity is to be less than the particular threshold at which the deposition of certain suspended particles starts. The main sources of the suspended sediment that accumulates in drains are the products of erosion, deflation and bed (including the slope slides and slumps) processes, as well as organic matter, presenting 36, 5 (the winter deflation products only), 13 and 20 % of the total, respectively (Lamsodis *et al.*, 2006). Herbs growing on the drain bed positively affect the silting rate, both by lowering flow velocity and by enhancing the organic matter content in sediment (Figure 13). As woody vegetation suppresses the helophytes, the drains having woody plants on their slopes experience the least bed-silting rates (Figure 14).

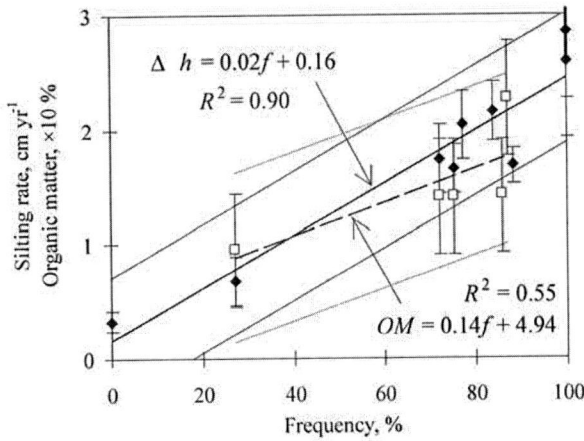

Figure 13. Dependence of sediment accumulation rate (Δh, rhombuses, thick solid trend-line) and sediment organic matter percentage (OM, squares, thick dotted trend-line) upon the frequency of occurrence of vegetation on the bed of a drain (f). The Δh and OM values are the means from 5–73 data series. The 'antennae' mark SE, both thin solid and dotted lines mark the confidence bonds of regressions at $p = 0.05$.

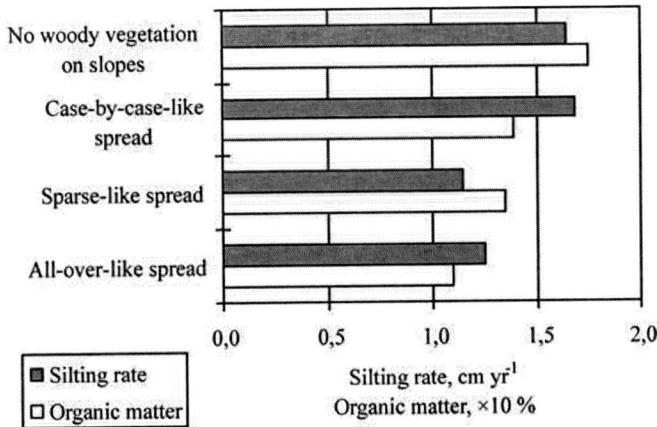

Figure 14. Sediment accumulation rate (Δh) and sediment organic matter percentage (OM) in drains exhibiting different character of spread of woody vegetation on slopes when the catchment area $A \geq 5.0$ km^2. The dependence of these indexes of sediment accumulation upon woody vegetation density on slopes (D, stems m^{-2}) shows the regression equations to be: $\Delta h = 1.60 - 0.30\,D$, r $= -0.71$ and $OM = 16.15 - 3.76\,D$, r $= -0.90$.

Example 6: Water Quality

Some alterations occur when water flows in from non-overgrown to overgrown, or from not dammed to beaver-impounded drain stretches. Those include the changes in bed shading and flow velocity that in turn influences the depth, insolation and temperature conditions of the water column, and the process of deposition of suspended sediment; in beaver ponds, all products of beaver activities accumulate too (Lavrov, 1981; Johnston and Naiman, 1987; Lamsodis, 2001). The hydrological and biogeochemical processes that are influenced by such alterations are responsible for nutrient cycling between organic and inorganic forms, *i.e.* re-mineralisation, microbially mediated transformations and biological uptake (Perelman, 1975; McClain *et al.*, 1998). In each case, the data collected demonstrates some retention of nutrients when water is flowing via overgrown or beaver impounded stretches of drains (Figures 15 and 16), resulting in the water quality improvement of a downstream runoff.

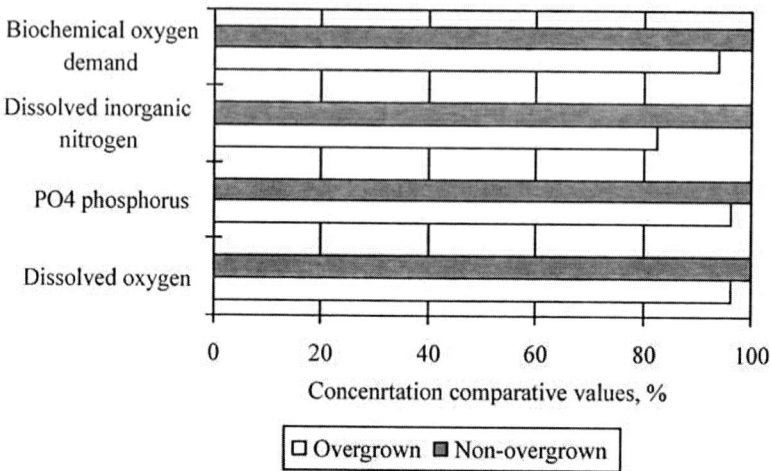

Figure 15. Alterations of mean yearly values of nutrient concentrations and the BOD_5 index when water was flowing in sequence via non-overgrown and overgrown stretches of drain. The differences between the concentrations in those variant stretches were not significant; however, the decreasing tendency in the mean concentration values demonstrated that water quality did not worsen in the vicinity of woody vegetation.

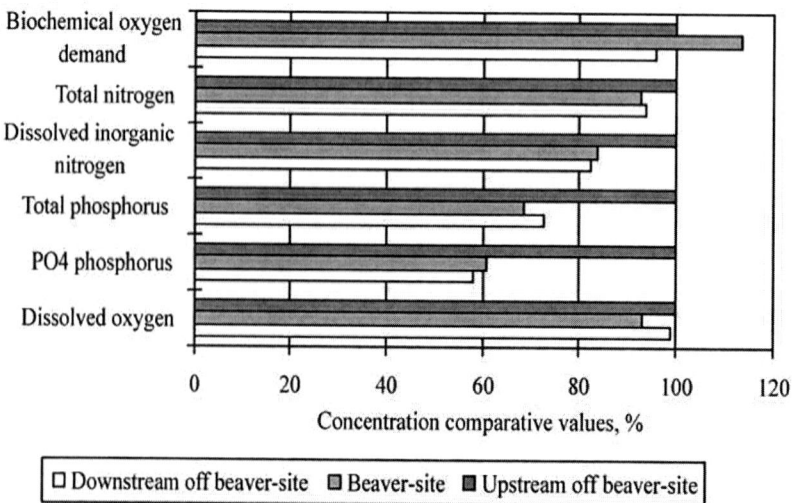

Figure 16. Alterations of mean yearly values of nutrient concentrations and the BOD_7 index when water was flowing via beaver impounded drains. The differences between concentrations of DIN, TP and PO_4-P occurred upstream off the beaver-site, and in the beaver-site were significant at $p < 0.01$.

CONSIDERATIONS CONCERNING THE EXAMPLES

In the context of the self-rehabilitation of a landscape, there are several points arising from each of the above examples, and also from the interaction of some of them.

1) The slope slumps and slides supply the drain and other downstream water-bodies with sediment and pollutants, and make the slopes difficult to maintain. However, slumps and slides progress the conversion of a drain's trapezium-shaped cross-section into one more resembling a parabola; such cross-sections are more stable than the man-made trapezium shapes. Although the drain bed has become wider, due to the cross-section deformations, and is thereby inclined to silting, it provides more room for a stream to meander which furthers the flow turbulence and carrying capacity of low gradient streams, thus preventing suspended sediment deposition even in the low-water periods. Conversely, in high gradient drains under the same hydrologic conditions, stream meandering slows the flow velocity, thus reducing the scouring efficiency of the flowing water and preventing drain bed incision. Hence, the meandering makes stream hydraulics to be more various. Clearly: (1) as a result of drain deformations, the diversity of the abiotic conditions within the drains increases (this is important for the environment); (2) both the cross-section stability increasing over time, and the possibility of meandering in the drain bed reducing sediment accumulation and bed incision, are useful in terms of drain management.

2) It is true to say that drains are almost the only places in the vast fields of today's tamed agro-landscape that can provide shelter for much of the wild plant and animal life. The decreasing number of species of woody vegetation in drains, the greater the distance they are from a forest, is the evidence of drains being ecological corridors even for vegetation. (There is no doubt that changing plant communities are followed by the emergence of other species of invertebrates; a greater diversity of bird and animal species can also be seen in such drains.)
The drain slopes that are completely overgrown with woody vegetation need no annual mowing. The roots of woody vegetation strengthen slopes and improve the filtration properties of the soil all year round (Molchanov, 1973, Danilov *et al.*, 1980). The decaying litter of trees and shrubs improves the structure of the topsoil and, along with the rooting, enhances the infiltration capacity of the slope soils. This makes the drain slopes more resistant to the erosion. As a result, (1) the succession of vegetation communities taking place in the drains can be seen to demonstrate the rehabilitation of a stream to an original function, such as giving refuge to various species, thus providing the chance for the heritage of genetic biodiversity to persist; (2) from a drain management viewpoint, drains become more stable when slopes are forested.

3) The enhanced infiltration capacity and surface roughness both provide the ability for the vegetation covering drain slopes and riparian buffer strips along the drains to remove suspended solids (erosion products) from surface runoff. Depending on the amount of runoff, buffer width and lateral gradient, those zones are capable to trap

66–96 % of total suspended solids when uniform overland sheet flow is maintained (Rachinskas, 1983).

Every shelterbelt provides a barrier against prevailing winds. If such shelter belts make up a net of green plantation bordering the fields (the forested drains might enter this net), a decrease in wind speed and turbulence may occur in the lowermost atmospheric strata (Konstantinov et al., 1965). Consequently, there is less threat of deflation originating in such bordered fields. Thereby, the features of forested slopes (and buffer strips) to trap erosion products as well as to mitigate the speed of wind that causes the deflation either (1) protect the drains from sediment and (2) preserve the adjacent soils from the physical denudation.

4) The increased infiltration in the overgrown strips accounts for a reduced amount of surface runoff that has passed over the strips and thereby the amount of dissolved pollutants (Vought et al., 1994). However, Rachinskas (1983) has revealed that during an erosion event most nutrients (about 90–95 %) were settled in the buffer strips, along with the interception of the eroded soil. Therefore, the vegetation retaining erosion products along with attached and dissolved nutrients may protect drains and downstream-located water-bodies from pollution. The study results indicate that there also was no impairment in the quality of that water which already was flowing in drains if slopes were overgrown by trees or shrubs.

5) Beavers act as very strong driving force transforming landscapes and drains with their environs therein. Ecological effect of beavers is mainly associated with the conversion of terrestrial ecosystems into aquatic ones with all subsequent consequences (Johnston et al., 1990; Hammerson, 1994; Collen et al., 2000) including the alterations in pathways and rates of nutrient cycling (Johnston et al., 1987; Naiman et al., 1994). Maret et al. (1987) has indicated the positive effect of beaver impoundments on the quality of running water. It would be emphasised that beavers essentially alter hydraulic conditions of obstructed drains: there are the stretches of uniform flow and riffles, pools and shoals changing together. The diversity of stream hydrologic conditions both in beaver obstructed drains and due to the above-mentioned meandering also contributes to the pollutant transformation process (Velner, 1976). Thus, beaver impoundments retaining nutrients that have entered into the drains upstream are able both (1) to control the downstream runoff pollution and (2 like buffer strips and forested slopes, preserve landscapes from the nutrient losses.

SUMMARY

All the benefits that issue from self-rehabilitation of drains are condensed in Table 3. These benefits concern the fields of both environmental and professional. Moreover, it would be noted that every alteration leads not only to the environmental recovery of drains but also assist in the maintenance of their functions to collect and discharge water from drainage systems. Just in this nature's suggestion the invitation lies to people to collaborate. Man both should recognize the use of bi-directional effects of the self-naturalisation processes in drains and their environs, and accept and utilise them for the improvement of the methods in drain management.

Table 3. Alterations, benefits and shortcomings occurring over time in drains and vicinity

Alteration	Alteration form	Environmental benefit	Professional benefit	Shortcoming
Long-term deformations of drain profiles	1) Parabola-shaped cross-section; (2) meandering	1) Diversity of abiotic conditions; (2) positive effect on water quality	1) Cross-section stability; (2) prevention of silting and bed incision	Some trouble when mowing
Long-term alterations of vegetation cover	1) Extinction and emergence of some herbaceous species; (2) emergence of woody vegetation	1) Diversity of species and communities; (2) prevention of water-bodies from pollution; (3) retention of nutrients in landscape; (4) soil conservation in case of shelterbelt net that prevents deflation	1) Stabilisation of slopes; (2) no mowing or reduced amounts of mowing; (3) no dragging or reduced amounts of dragging; (4) lessened thread for topsoil degradation due to deflation; (5) wood	1) Some trouble when mowing; (2) manual work and/or special equipment and means are necessary if need in dragging would occur; (3) decrease in hydraulic capacity; (4) potential of roots to penetrate into tile drain tubing
Hydrographical changes	Beaver impoundments	1) Diversity of hydraulic conditions; (2) diversity of habitable conditions; (3) diversity of species, communities and ecosystems; (4) retention of water; (5) accumulation of suspended sediment; (6) water quality improvement; (7) retention of nutrients in landscape; (8) soil conservation due to raised erosion basis	1) Lessened soil losses due to erosion prevention; (2) water reservoirs in case of fire; (3) game	1) Headed-up drainage systems; (2) sometimes occurring inundation of adjacent fields; (3) damage to wood plantations; (4) damage to embankments, dikes and other structures; (5) some damage to crop

In order to avoid and/or mitigate shortcomings of self-naturalisation there are different methods proposed how to control the hydraulic capacity of afforested drains (Rimkus *et al.*, 2003) as well as water level in beaver ponds including restoration of drainage in headed tile systems (Laramie, 1963; Heidecke, 1985; Lamsodis, 2003), damage to road-beds, culverts, dams, riparian dikes and trees (Laramie, 1963; Heidecke, 1985; Schrenk, 1996). Analysing the territorial aspects of the development processes in landscapes, Tiknius (2002), however, warned people that their actions have to release and actuate the self-development potencies of nature therefore, the abrupt intrusion and methods of direct management of landscape should be an exception, not a rule. In this viewpoint, every maintenance work and particularly the policy of management of drains should sustain development their original functions. The efforts of man to maintain the designed parameters and thus to further the inflow and outflow functions of drains should encourage the processes of re-naturalisation. In particular, this concerns those of them that lead to 'self-management' of drains. It will be observed that some suggestions how drains would have to be maintained and reconditioned for their management to become ecologically reasonable are currently being made (Lamsodis *et al.*, 2006). Tiknius (2002), incidentally, has also mentioned, "The original landscape owns the worth that nothing can substitute. ... Consequently, not every well-meant and scientifically based reorganisation of landscape makes a sense: it is vital to save certain part of any emergent and progressing structures in landscape despite the level at which it is urbanized."

REFERENCES

Agriculture and forestry (2000). In *Statistical Yearbook of Lithuania 2000* (pp. 379–429). Vilnius, LT: VĮ Statistikos centras.

Basalykas, A. (1977). *Lietuvos TSR kraštovaizdis* [Landscape of Lithuanian SSR]. Vilnius, SU: Mokslas (in Lithuanian).

Bastian, O., & Bernhardt, A. (1993). Anthropogenic landscape changes in Central Europe and the role of bioindication. *Landscape Ecology, 8*, 139–151.

Berankienė, L. (1997). A vegetation cover research on at land-reclamation canal slopes and bottom. *Vandens ūkio inžinerija: mokslo darbai, 3(25)*, 178–183 (in Lithuanian).

Collen, P., & Gibson, R. J. (2000). The general ecology of beavers (Castor spp.) as related to their influence on stream ecosystems and riparian habitats, and the subsequent effects on fish – a review. *Reviews in Fish Biology and Fisheries, 10*, 439–461.

Danilov, G. G., Lobanov, D. A., & Kargin, I. F. (1980). *Effektivnost agrolesomelioratsii v Nechernoziomnoy zone RSFSR* [Effectiveness of Agricultural Afforestation in Non-Chernizem Zone of R.S.F.S.R.]. Moskva, SU: Lesnaya promishlennost (in Russian).

Décamps, H., Pinay, G., Naiman, R. J., Petts, G, E., McClain, M. E., Hillbrcht-Ilkowska, A., Hanley, T A., Holmes, R. M., Quinn, J., Gibert, J., Planty Tabacchi, A-M., Schiemer, F., Tabacchi, E., & Zalewski, M. (2004). Riparian zones: Where biogeochemistry meets biodiversity in management practice. *Polish Journal of Ecology, 52*, 3–18.

Dilys, A., & Minkevičius, V. (1997). Winter deflation of soils in the West Lithuania. *Geografijos metraštis, 30*, 191–198 (in Lithuanian).

Eitminavichiene, N. (1984). Peculiarities of landscape-differentiated territory of Lithuania in the 16th century. *Geographia Lituanica, 21*, 174–179 (in Russian).

Falk, K. J., Burke D. M., Elliott, K. A., & Holmes, S. B. (2008). Effect of single-tree and group selection harvestingon the diversity and abundance of spring fores herbs in deciduous forests in southwestern Ontario. *Ecology and Management, 255*, 2486–2494.

Forman, R. T. T. (1995a). Some general principles of landscape and regional ecology. *Landscape Ecology, 10*, 133–142.

Forman, R. T. T. (1995b). *Land Mosaics: The Ecology of Landscapes and Regions*. Cambridge, UK: Cambridge University Press.

Hammerson, G. A. (1994). Beaver (Castor canadiensis): Ecosystem alterations, management, and monitoring. *Natural Areas Journal, 14*, 44–57.

Heidecke, D. (1985). Ergebnisse und Probleme beim Schutz des Elbebibers. *Naturschutz in Berlin u. Brandenburg, 21*, 6–12.

Hietala-Koivu, R., Lankoski, J., & Tarmi, S. (2004). Loss of biodiversity and its social cost in an agricultural landscape. *Agriculture, Ecosystems and Environment, 103*, 75–83.

Johnston, C. A., & Naiman, R. J. (1987). Boundary dynamics at the aquatic-terrestrial interface: The influence of beaver and geomorphology. *Landscape Ecology, 1*, 47–57.

Kaušyla, K. (1981). Climate and its resources. In VGKV gamybinis žemėlapių sudarymo susivienijimas „Kartografija" (Eed.), *Lietuvos TSR atlasas* (p. 58). Maskva, SU: Vyriausioji geodezijos ir kartografijos valdyba prie TSRS Ministrų Tarybos (in Lithuanian).

Konstantinov, A. P., & Struzer, L. P. (1965). *Lesnie polosi i urozhay* [Forest Shelter Belts and Crop]. Leningrad, SU: GIMIZ (in Russian).

Lamsodis, R. (2003). Beavers in drainage channels: an attempt to control the consequences. In: Latvia University of Agriculture (ed.), *Research for Rural Development 2003*. International Scientific Conference Proceedings, Jelgava, Latvia, 21-24 May, 2003 (pp. 87–93), Jelgava, Latvia (in Russian).

Lamsodis, R. (2008). Impact of beavers inhabiting open drains on runoff. In Lietuvos žemės ūkio universiteto Vandens ūkio institutes (Ed.), *2007 m. mokslo tiriamieji darbai* (pp. 47–49). Vilainiai, LT: UA „Rinkos aikštė" (in Lithuanian).

Lamsodis, R. (2008). Geomorphologic alterations in beavers inhabited open drains. In Lietuvos žemės ūkio universiteto Vandens ūkio institutes (Ed.), *2007 m. mokslo tiriamieji darbai* (pp. 52–53). Vilainiai, LT: UAB „Rinkos aikštė" (in Lithuanian).

Lamsodis, R., Morkūnas, V., Poškus, V., & Povilaitis, A. (2006). Ecological approach to management of open drains. *Irrigation and Drainage*, 55, 479–490.

Lamsodis, R., & Poškus, V. (2006). Vegetation in drainage channels: interface with sediment accumulation. *Water Management Engineering, 3(6)*, 60–68.

Lamsodis, R., Poškus, V. Ragauskas, S. (2006). Sediment accumulation in drainage channels: Sources, rate of silting, composition. In A. Dumbrauskas, F. Mikuckis, & P. Punys (Eds.), *Advanced Methods and Solutions in Water Engineering*. Selected papers of the conference, dedicated to the 60 year anniversary of the Faculty of Water and Land Management, Lithuanian University of Agriculture, 12–13 October 2006, Birštonas, Lithuania (pp. 81–84) (in Lithuanian).

Laramie, H. A. (1963). A Device for Control of Problem Beavers. *Journal of Wildlife Management, 27*, 471–475.

Lasinskas, M. (1981). Surface waters and their resources. In: VGKV gamybinis žemėlapių sudarymo susivienijimas „Kartografija" (ed.), *Lietuvos TSR atlasas* (p. 78). Maskva, SU: Vyriausioji geodezijos ir kartografijos valdyba prie TSRS Ministrų Tarybos (in Lithuanian).

Lavrov, L. S. (1981). *Bobri Paleoarktiki* [Beavers of Palearctic]. Voronezh, SU: Izd. Voronezhskogo universiteta (in Russian).

Maret, T. J., Parker, M., & Fannin, T. E. (1987). The effect of beaver ponds on the nonpoint source water quality of a stream in Southwestern Wyoming. *Water Research, 21*, 263–268.

Matulionis, P. (1930). Land and population of Lithuania in the course of history. *Kultūra, 5*, (in Lithuanian).

McClain, M. E., Bilby, R. E., & Triska, F. J. (1998). Nutrient Cycles and Responses to Disturbance. In R. J. Naiman, & R. E. Bilby (Eds.), *River Ecology and Management: Lessons from the Pacific Coastal Ecoregion* (pp. 347–372). New York, NY: Springer.

Molchanov, A. A. (1973). *Vliyanie lesa na okruzhayushchuyu sredu* [Impact of Forest on the Environment]. Moskva, SU: Nauka (in Russian).

Müller, Ch., Berger, G., & Glemnitz, M. (2004). Quantifying geomorphological heterogeneity to assess species diversity of set-aside arable land. *Agriculture, Ecosystems and Environment, 104*, 587–594.

Naiman, R. J., Décamps, H., & Pollock, M. (1993). The role of riparian corridors in maintaining regional biodiversity. *Ecological Applications, 3*, 209–212.

Naiman, R. J., & Pinay, G., Johnston, C. A., & Pastor, J. (1994). Beaver influences on the long-term biogeochemical characteristics of boreal forests drainage networks. *Ecology, 75*. 905–921.

Palang, H., Mander, Ü., & Luud, A. (1998). Landscape diversity changes in Estonia. *Landscape and Urban Planning, 41*, 163–169.

Paulyukyavichyus, G., & Yuodis, Yu. (1987). Resistance of agrolandscapes to the anthropogenic press: Antidenudational resistance of landscapes: Estimation of lands in danger of deflation. In V. E. Sokolov, T. V. Vasileva, V. L. Kontrimavichus, R. Yu. Pakalnis, & G. B. Paulyukavichus (Eds.), *Ekologicheskaya optimizatsiya agrolandshafta* (pp. 71–78). Moskva, SU: Nauka (in Russian).

Perelman, A. I. (1975). *Geochemistry of Landscape*. Moskva, SU: Visshaya shkola (in Russian).

Pielou, E. (1975). *Ecological Diversity*. New York, NY: Wiley Interscience.

Povilaitis, A. (2000). Changes in the structure of rural landscape and drainage canals network in Lithuania. In *Land Reclamation and Landscape Management*. Proceedings of International Conference, 8–9 June 2000, Kaunas–Akademija, Lithuania (pp. 104–108) (in Lithuanian).

Povilaitis, A. (2001). Historical and territorial peculiarities of Lithuanian agricultural landscape. *Vagos, 52 (5)*, 62–72 (in Lithuanian).

Rachinskas, A. (1983). Substantiation and method for determination of reasonable width of riparian buffer belts. In *Trudi Litovskogo nauchno-issledovatelskogo instituta gidrotekhniki i melioratsii i Litovskoy ordena trudovogo krasnogo znameni selskokhozyayctvennoy akademii: Okhrana vodoistochnikov i ratsionalnoe ikh ispolzovanie* (vol. 14, pp.82–98). Vilnius, LT: Mokslas (in Russian).

Rachinskas, A., & Belyauskas, P. (1987). Resistance of agrolandscapes to the anthropogenic press: Antidenudational resistance of landscapes: Antierosional resistance. In V. E. Sokolov, T. V. Vasileva, V. L. Kontrimavichus, R. Yu. Pakalnis, & G. B. Paulyukavichus (Eds.), *Ekologicheskaya optimizatsiya agrolandshafta* (pp. 62–71). Moskva, SU: Nauka (in Russian).

Račinskas, A., & Morkūnaitė, R. (1986). Deflation in Lithuania. *Žemės ūkis, 12,* 13 (in Lithuanian).

Ragauskas, S. (2002). Investigations on the deterioration of slopes in the process of mechanized maintenance of channels. *Vandens ūkio inžinerija, 20(42)*, 82–87 (in Lithuanian).

Rimkus, A., Lamsodis, R., & Vaikasas, S. (2003). Naturalization of drainage channels in Lithuania and possibilities of their maintenance as water recipients. *Nordic Hydrology, 34*, 493–506.

Rosell, F., Bozsér, O., Collen, P., & Parker, H. (2005). Ecological impact of beavers *Castor fiber* and *Castor Canadensis* and their ability to modify ecosystems. *Mammal Review, 35*, 248–276.

Schlosser, I. J, & Karr, J. R. (1981). Water quality in agricultural watersheds: Impact of riparian vegetation during base flow. *Water Resources Bulletin, 17*, 233–240.

Schoonover, J. E., Williard, K. W. J., Zaczek, J. J., Mangun, J. C., & Carver, A. D. (2006). Agricultural sediment reduction by giant cane and forest riparian buffers. *Water, Air, and Soil Pollutoin, 169*, 303–315.

Schrenk, G. (1996). Der Biber – Wasserwirtschaftliche Probleme und Lösungsmöglichkeiten. *Wasser & Boden (DVWK: gn-info), 48*, 22–24.

Tiknius, A. (2002). *Teritoriniai vystymosi bruožai* [The Territorial Aspects of the Development Processes]. Klaipėda, LT: Klaipėdos universiteto leidykla (in Lithuanian).

Vaičys, M., Armolaitis, K., Raguotis, A., Kubertavičienė, L., Matusevičius, K., Lubytė, J., Eitminavičius, L., Šleinys, R., Janušienė, V., Rimšelis, J., Čeburnis, D., Kvietkus, K., Eitminavičiūtė, J., Bagdanavičienė, Z., Budavičienė, I., Strazdienė, V., Pauliukevičius, G., Gulbinas, Z., Morkūnaitė, R., Jankauskaitė, M., Baubinas, R., Dilys, A., Aleknavičius, P., Jasinskas, J., Kavaliauskas, P., & Dumbliauskienė, M. (1998). Land use, soil pollutedness and ways for ecology sustaining maintenance. In Tarptautinio Mokslinės Kultūros Centro – Pasaulinės Laboratorijos Lietuvos skyrius (Ed.), *Regijono ekologinis tvarumas istoriniame kontekste* (pp. 40–55). Vilnius, LT: Mokslo aidai (in Lithuanian).

Velner, Ch. A. (1976). Processes of pollutant transformation in water courses. In *Rechnaya gidravlika i ruslovie protsessi* [River Hydraulics and Bed Processes] (vol. I, pp. 124–132). Moskva, SU: Izd. Moskovskogo universiteta (in Russian).

Vought, L. B. M., Hahl, J., Pedersen, C. L., & Lacoursiere, J. O. (1994). Nutrient retention in riparian ecotones. *Ambio, 23*. 342–348.

In: Ecological Restoration
Editors: George H. Pardue and Thomas K. Olvera

ISBN 978-1-60741-013-3
© 2009 Nova Science Publishers, Inc.

Chapter 8

Evaluation of Biological Quality of Lotic Ecosystems in Central-Southern Italy: A Comparison of Different European Biotic Indices

Laura Mancini[1,1], Serena Bernabei[2], Carlo Jacomini[2], Valentina Della Bella[1], Stefania Marcheggiani[1] and Lorenzo Tancioni[3]

[1]Department of Environment and Primary Prevention,
National Institute of Health, Rome, Italy
[2]Agency for Environmental Protection and Technical Services
(APAT) of Itay, Via Vitaliano Brancati, Rome, Italy
[3]Laboratory of Experimental Ecology and Aquaculture (L.E.S.A),
Biology Department, University of Rome Tor Vergata, Rome, Italy

Abstract

In Italy, rivers and streams show very different typologies. Our analysis of lotic ecosystems in Central and Southern Italy show up how different river typologies may alter the evaluations of biological quality. We monitored eight rivers with particular features and located in different Italian regions. Three European biological methods, all using the macro-invertebrate communities as biological indicators of water quality, have been compared from these rivers data sets. In particular, the *Indice Biotico Esteso* (IBE; Italy), Belgian Biotic Index (BBI; Belgium), and Biological Monitoring Working Party (BMW.P; Spain) were compared. We accomplished qualitative and quantitative samples in order to calculate index values. IBE and BBI use both the indicator taxa and the total sum of taxa in the community as common basis to reckon the index value, while BMWP scores a value for all sampled groups, determined at the family level. All these three methods transform the index values into Quality Classes, whose ranking is not so different from the Index Value. On spatial scale, comparisons were performed (where

---
[1] Corrensponding author: tel. +39 06 49902773, fax: +39 06 2861, e-mail: laura.mancini@iss.it;

possible) within each river and among different rivers. Our results seem to suggest that IBE is a quite well fitted index, and its authors are now working to improve its fitting to the peculiar conditions of Southern Italy streams. BBI instead gives excessive values due to its fitting to different river typologies. BWMP shows a trend as to increase the obtained values but it gives an acceptable value to the whole community to evaluate the biological quality of waters.

Keywords: macroinvertebrates, river quality, biotic indices, IBE, EBI, BWMP

1. INTRODUCTION

The definition of biological monitoring methodologies needs extensive settings and refinements in order to accomplish the objectives required by European Water Framework Directive 2000/60/CE (CEC, 2000), particularly in running water if different typologies of rivers are concerned (Mancini, 2005).

A tentative approach to highlight possible differences and affinities among different rivers in Central-Southern Italy has been accomplished to review the knowledge on river quality. This is of major concern, as most of the studies and researches in Italy have been performed in the North, while there is a lack of extensive research on the water quality of most of the Southern Italy streams. A review of the situation in Southern Italy (Mancini *et al.*, 1995b) showed some of the main differences in those river typologies. In this paper, we want to review some of those rivers, and propose a comparison of three European biotic indices applicable to the available data sets.

2. MATERIALS AND METHODS

2.1. Examined Rivers

We monitored eight rivers of Central-Southern Italy with particular features (Figure 1). The Albegna River is a Tuscanian pristine stream (68.5 km long, 748.6 km^2 of hydrological basin), receiving a thermo-sulphurated tributary in its mid-course. It runs through an unindustrialised country, showing a steep gradient from the alpine-like upstream to the agricultural lowlands downstream.

The Arrone River (35 km long, 275 km^2 of hydrographic basin, with 34,000 inhabitants), outpours from Bracciano Lake, in Latium. It receives the waters from two sewage treatment plants, and as the previous one runs through broadleaf woods of high naturality and agricultural lands.

The Tiber River, third Italian river in length (405 km) and second in surface of hydrological basin (17,156 km^2), crosses 12 Province Districts in 4 Regions, resulting the most important river of Peninsular Italy. The Total Equivalent Inhabitants (TEI) are 10,283,187. After a mid-course affected by agro-zootechnical activities, its downstream is severely polluted by partially-depurated civil waste waters (lastly, Rome) and industrial discharges.

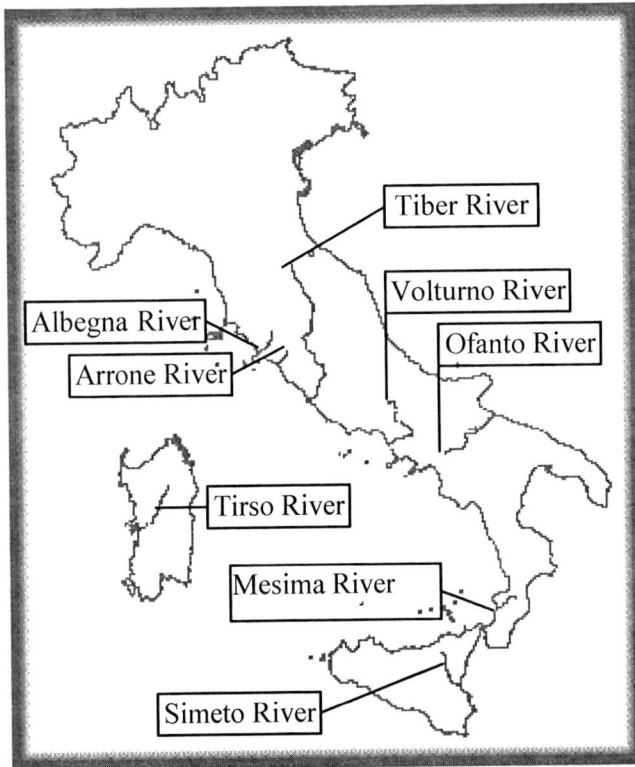

Figure 1. Italy: Map of the examined rivers.

The Volturno River is the main Campanian river (175 km), with a particularly branched hydrographic basin of 5,680 km^2, passing through very different anthropogenic disturbances. The Total Equivalent Inhabitants (TEI) are 1,516,361. There are many urban areas along its course, and it is considered a fragile and riskful system for the aquatic life as it receives pressures by excavation plants, and civil, farm and industrial areas. It is severely polluted.

The Mesima River (707 km^2) is a Calabrian system passing in its mountainous upstream through scarcely populated areas, while on the floodplains of the mid- and lower course it receives the waste waters of several urban areas, the biggest of which is the town of Rosarno, on the estuary. The Mesima river is polluted by uncontrolled discharges, mainly agricultural and agro-industrial ones.

The Ofanto River (165 km, 2,780 km^2) crosses Campania, Basilicata and Apulia, where 175,998 people are living, with a human impact more than double, corresponding to 362,533 TEI. Waters outpouring from its spring are already labelled as unpotable. The pollution sources are heavy industry, intensive agriculture, undepurated civil sewages. Twenty years ago it resulted moderately polluted; nowadays it is classified as heavily polluted.

The Tirso is the longest river in Sardinia (159 km), covering an area of 3,376 km^2 with 164,510 resident inhabitants and 481,956 Total Equivalent Inhabitants. Farms and urban areas discharge in it their waste waters, often without any depuration, a situation worsened by some industrial sludges occurring in its mid course. Besides this, a dam forms the Omodeo Lake, after which other smaller man-made barrages work as sedimentation basins, although interrupting the river biological continuum.

The Simeto River is the second longest Sicilian river (116 km), with the widest basin (4,186 km^2, 436,977 resident inhabitants, 703,576 TEI). While crossing four Province Districts, it receives several animal and civil waste waters. In the terminal tract, the pollution levels are concentrated due to the water caption for agricultural uses. Its average reach is of 25 m^3/s, with a range from 1 in summer to 1,500 m^3/s in winter, so its basin is characterised by heavy hydrological problems (e.g. landslides, summer droughts), mainly due to the lowest wood cover (about the 3% of the total basin area) and to the dams and barrages built up for different human activities, that sometimes make some river tracts to dry up.

2.3. Field Sampling and Biotic Indices

The study rivers were seasonally sampled between 1997 and 2000. The total number of available samples were 146. Macroinvertebrates were collected with a hand net (21 mesh per cm^2) and, after a first sorting on-site, were identified in the laboratory.

Biological monitoring is a tool currently used to control lotic (= running-waters) ecosystem quality. In particular, the study of river populations and communities allows to get quality indices and maps utilised in every evaluation of either the "naturality" degree or the pollution level, and to obtain immediately information useful for land management.

The basis of this principle is that each river gathers within its waters the leaching waters and thus the pollutants discharged in the hydrological catchment. Such pollutants may result extremely diluted, so much so that they may be difficult to detect by simple chemical-physical analyses, meanwhile they affect profoundly the structure and the composition of natural communities. Organisms and populations living in the river are affected by the pressure of pollutants (here considering every alteration of the pristine environment, included temperature, pH, turbidity, etc.), and vary their numbers in specimens and species composition according to the conditions of the water they live in. Therefore, a modification of the environment which could be undetected by abiotic methods may instead be highlighted by means of a simple analysis of its biological community.

Macroinvertebrates from the river bottom (macro-zoobenthos) are the most used pool of these organisms, as they can be placed at an intermediate level in the food web, representing the link between higher (e.g., fish) and lower (e.g., micro-decomposers) levels, are quite easy to collect and to identify, and most of their species are quite well studied, so much so that it is possible to know both autoecological and sinecological aspects of their natural history. Some taxa result very sensitive to pollutants, resulting in good indicators of water quality, while others are extremely resistant.

Biological indices and indicators elaborated in the last decades in Europe may be gathered within three main categories: Saprobic Indices, based on indicator species of different levels of water quality, Diversity Indices, based on species diversity, and finally Biotic Indices, combining the indicator value of some species with the species richness of the community (Ghetti, 1997).

We examined several different rivers of Central-Southern Italy, whose data sets had been obtained by standard biological methodologies to evaluate the water quality.

The *Indice Biotico Esteso* (thereafter named IBE) (Ghetti, 1997, 1995) is an updated and revised version of the former Extended Biotic Index (EBI, Ghetti, 1986). Its structure derives from the Indices elaborated by Woodwiss (1964, 1978, 1980), later adapted to the Italian environments (Ghetti, 1986, 1995, 1997, Ghetti & Bonazzi, 1980).

The first step to reckon the IBE value is the analysis of the community structure, through qualitative samples of aquatic macro-invertebrates, determined in the lab at the systematic levels of family or genus (Tab. 1).

Table 1. Limits to define the Systematic Units (S.U.)

Faunal Groups	Determination Levels
Plecoptera	genus
Trichoptera	family
Ephemeroptera	genus
Coleoptera	family
Odonata	genus
Diptera	family
Eteroptera	family
Crustacea	family
Gastropoda	family
Bivalvia	family
Tricladia	genus
Hirudinea	genus
Oligochaeta	family
Other taxa to consider in order to reckon the IBE	
Sialidae (Megaloptera)	
Osmylidae (Planipennia)	
Prostoma (Nemertina)	
Gordidae (Nematomorpha)	

The IBE is then calculated according to a double input table (Tab. 2): in row 1 the numbers of systematic units (S.U.) present are ranked; in column 1 the sensitivity of observed faunal groups is listed, ranked from 1 to 7 with regard to decreasing environmental requirements or increasing tolerance to pollution. Crossing the selected row and column, the value of the index is easily obtained for the sampling station considered.

The Index values may then be converted into Quality Classes (Tab. 3), standardised during several European Technical Seminars, which allow to translate the obtained index values into environmental evaluations, and also to plot them with colours onto a map of the river. This results in a easy way to depict the results, and results useful also for managers and unspecialised readers.

Similar indices are used in several Countries in our Continent (Tab. 4), and among them we selected the Belgian Biotic Index (BBI) (De Pauw & Vanhooren, 1983, De Pauw, 1988), and the Biological Monitoring Working Party (BMWP) (Hellawell, 1978) modified for Spanish rivers (Alba-Tercedor & Sánchez-Ortega, 1988). BBI and BMWP, as well as the IBE, translate their results in Quality Classes, thus they may be comparable and subject to statistical inference.

Table 2. Table to reckon the IBE value.

Presence of taxa in the catch (first entry)		Amount of taxa in the catch (second entry)								
		0-1	2-5	6-10	11-15	16-20	21-25	26-30	31-35	>36
1. Plecoptera (*Leuctra*)[b]	>1 genus	-	-	8	9	10	11	12	13[a]	14[a]
	1 genus	-	-	7	8	9	10	11	12	13[a]
2. Ephemeroptera (excluding Baetidae & Caenidae)[c]	>1 genus	-	-	7	8	9	10	11	12	-
	1 genus	-	-	6	7	8	9	10	11	-
3. Trichoptera (Baetidae and Caenidae included)	>1 family	-	5	6	7	8	9	10	11	-
	1 family	-	4	5	6	7	8	9	10	-
4. Gammaridae and/or Atiidae, and/or Palaemonidae	the above taxa missing	-	4	5	6	7	8	9	10	-
5. Asellidae and/or Niphargiidae	the above taxa missing	-	3	4	5	6	7	8	9	-
6. Oligochaeta or Chironomiidae	the above taxa missing	1	2	3	4	5	-	-	-	-
7. Other organisms	the above taxa missing	-	-	-	-	-	-	-	-	-

a : Rarely occurring in Italy; necessary to be aware not to sum up biotypologies (artificial increase in taxa richness).

b : In a community where only *Leuctra* is present as Plecoptera and Ephemeroptera are lacking (or only Baetidae and Caenidae are present), *Leuctra* must be considered as Trichoptera.

c: Baetidae and Caenidae families must be considered at the Trichoptera level for the horizontal entry.

-: Uncertain results due to wrong sampling, drift of benthos fauna, improper colonization and river typology out of IBE conditions (springs, snow melting waters, stagnant waters, estuaries and brackish zones).

Table 3. Conversion Table from IBE into Quality Classes and colours.

Quality Class	IBE Value	Evaluation	Colour
I	10-11-12-...	Slightly or unpolluted	blue
II	8-9	Slightly polluted	green
III	6-7	Moderately polluted – critical situation	yellow
IV	4-5	Heavily polluted	orange
V	0-1-2-3	Very heavily polluted	red

Table 4. Application of major index methods for assessment of running waters in EC Countries based on macroinvertebrates (modif. from: De Pauw & Hawkes, 1993)

Country	Index Method	Sampling[1]	Analysis[1]	Identification[2]	Standard[3]	Range
BELGIUM	BBI	QL	QL	O F G	N	0-10
DENMARK	Dansk Fauna Ind.	QL	QL	F S G	N	1-4
FRANCE	IBG	QL /QT	QL	F	N	0-20
GERMANY	BEOL/S	QL	QT	S	N	0-100/1-4
GREECE	-	-	-	-	-	-
IRELAND	Q-rating	QL	QL	F S G	N	0-5
ITALY	IBE	QL	QL	O F G	N	0-14
LUXEMBOURG	IB	QL	QL	O F	N	0-10
NETHERLANDS	K135	QL	QL	F S G	R	100-500
PORTUGAL	BBI	QL	QL	O F G	-	0-10
SPAIN	BMWP'	QL	QL	F	N	0->150
UK	BMWP/ASPT	QL	QL	F	N	0->150/0-10

[1] : QL = Qualitative Sampling; QT = Quantitative Sampling.
[2] Required identification level: O= Order; F= Family; G= Genus; S= Species.
[3] : N = National; R = Regional.

The BBI was elaborated according to the strategies of two previous indices, the Trent Biotic Index of Woodwiss (1964) and the Biotic Index of Tuffery & Verneaux (1968). The macro-invertebrate community structure is determined at several systematic levels (family, genus, presence of higher taxa in the case of Hydracarina) (Table 5).

Table 5. Practical limits to determinate Systematic Units for BBI

Taxonomic group	Determination Level of Systematic Units
Plathelminthes	genus
Oligochaeta	family
Hirudinea	genus
Mollusca	genus
Crustacea	family
Plecoptera	genus
Ephemeroptera	genus
Trichoptera	family
Odonata	genus
Megaloptera	genus
Hemiptera	genus
Coleoptera	family
Diptera	family
	Chironomiidae *thummi-plumosus*
	Chironomiidae non *thummi-plumosus*
Hydracarina	presence

The Belgian Biotic Index is calculated according to a standard table with double input (Tab. 6): in row 1 are ranked the numbers of S.U. present, and in column 1 the sensitivity of observed faunal groups, ranked from 1 to 7 with regard to decreasing environmental requirements or increasing tolerance to pollution.

Table 6. Table to reckon the BBI value.

I	II	III				
Faunistic Group		**Total Numbers of Systematic Units Present**				
		0-1	2-5	6-10	11-15	16 and more
				Biotic Index		
1. Plecoptera or EphemeropteraEcdyonuridae (= Heptageniidae)	1. several S.U.	-	7	8	9	10
	2. only 1 S.U.	5	6	7	8	9
2. Cased Trichoptera	1. several S.U.	-	6	7	8	9
	2. only 1 S.U.	5	5	6	7	8
3. Ancylidae or Ephemeroptera except Ecdyonuridae	1. more than 2 S.U.	-	5	6	7	8
	2. 2 or more S.U.	3	4	5	6	7
4. *Aphelocheirus* or Odonata or Gammaridae or Mollusca (except Spheriidae)	all S.U. mentioned above are absent	3	4	5	6	7
5. Asellus or Hirudinea or Spheriidae or Hemiptera (except *Aphelocheirus*)	all S.U. mentioned above are absent	2	3	4	5	-
6. Tubificidae or Chironomiidae of the *thummi-plumosus* group	all S.U. mentioned above are absent	1	2	3	-	-
7. Eristalinae (= Syrphidae)	all S.U. mentioned above are absent	0	1	1	-	-

Crossing the selected row and column, the value of the index is easily obtained for the sampling station examined. As for the IBE, the BBI can be converted into five Quality Classes, which can be plotted as different colours on maps (Tab. 7).

The BMWP instead reckons a score (from 1 to 10) for each taxon determined at the family level (Tab. 8), and the total sum of the scores collected determines the value of the index, which allows to classify the water quality in six (5+1) Quality Classes (Tab. 9).

Table 7. Conversion Table from BBI into Quality Classes and colours.

Class	Biotic Index	Colour	Significance
I	10-9	blue	slightly or unpolluted
II	8-7	green	slightly polluted
III	6-5	yellow	moderately polluted - critical situation
IV	4-3	orange	heavily polluted
V	2-0	red	very heavily polluted

Table 8. BMWP Score Table

Families	Values
Siphlonuridae, Heptageniidae, Leptophlebiidae, Potamanthidae, Ephemeridae, Taeniopterygidae, Leuctridae, Capniidae, Perlodidae, Perlidae, Chloroperlidae, Aphelocheiridae, Phryganeidae, Molannidae, Beraeidae, Odontoceridae, Leptoceridae, Goeridae, Lepidostomatidae, Brachycentridae, Sericostomatidae, Athericidae, Blephariceridae	10
Astacidae, Lestidae, Calopterygidae, Gomphidae, Cordulegasteridae, Aeschnidae, Corduliidae, Libellulidae, Psycomyidae, Philopotamidae, Glossosomatidae	8
Ephemerellidae, Nemouridae, Rhyacophilidae, Polycentropodidae, Limnephilidae	7
Neritidae, Viviparidae, Ancylidae, Hydroptilidae, Unionidae, Corophiidae, Gammaridae, Plactynemididae, Coenagrionidae	6
Oligoneuriidae, Dryopidae, Elmidae, Helophoridae, Hydrochidae, Hydreanidae, Clambidae, Hydropsychidae, Tipulidae, Simuliidae, Planariidae, Dendrocoelidae, Dugesiidae	5
Baetidae, Caenidae, Haliplidae, Curculioniidae, Chrysomaelidae, Tabanidae, Stratiomyidae, Empididae, Dolichopodidae, Dixidae, Ceratopogonidae, Anthomyidae, Limoniidae, Psychodidae, Sialidae, Piscicolidae, Hydracarina	4
Mesoveliidae, Hydrometridae, Gerridae, Nepidae, Naucoridae, Pleidae, Notonectidae, Corixidae, Helodidae, Hydrophilidae, Hygrobiidae, Dytiscidae, Gyrinidae, Valvatidae, Hydrobiidae, Lymnaeidae, Physidae, Planorbidae, Bithyniidae, Bithinellidae, Sphaeriidae, Glossiphoniidae, Hirudinidae, Erpobdellidae, Asellidae, Ostracoda	3
Chironomiidae, Culicidae, Muscidae, Thaumaleidae, Ephydridae	2
Oligochaeta (all the classes)	1

Table 9. Conversion Table from BMWP into Quality Classes and colours.

Class	BMWP	Evaluation of waters	Colour
	>150	Excellent quality	
I	101-150	Unpolluted	blue
II	61-100	Evident some effects of pollution	green
III	36-60	Polluted	yellow
IV	16-35	Heavily polluted	orange
V	<15	Very heavily polluted	red

2. RESULTS

The descriptive statistics for the three indices and their related Quality Classes are shown in Tab. 10.

Table 10. Descriptive statistics for IBE, BBI, BMWP and their Quality Classes.

	Valid N.	Mean	Confidence -0.95	Confidence 0.95	Sum	Minimum	Maximum	Var.Int.
IBE	146	7.84	7.43	8.24	1144	0	13	13
QC IBE	146	2.35	2.17	2.53	343	1	5	4
BMWP	146	69.36	62.53	76.19	10127	0	210	210
QC BMWP	146	2.52	2.31	2.73	368	1	5	4
BBI	146	7.77	7.41	8.14	1135	0	10	10
QC BBI	146	1.86	1.69	2.04	272	1	5	4

	Variance	Std.Dev.	St. Error	Asymm.	Std.Err. Asymm.	Kurtosis	Std.Err. Kurtosis
IBE	6.26	2.50	0.21	-0.58	0.20	0.55	0.40
QC IBE	1.24	1.11	0.09	0.55	0.20	-0.32	0.40
BMWP	1744.33	41.77	3.46	0.52	0.20	-0.07	0.40
QC BMWP	1.59	1.26	0.10	0.47	0.20	-0.87	0.40
BBI	4.89	2.21	0.18	-1.23	0.20	1.44	0.40
QC BBI	1.15	1.07	0.09	1.16	0.20	0.69	0.40

As an applied example of the Italian Index, in Figure 2 we show the IBE values for one of the examined rivers, the pristine Albegna River. It is clear how both on a spatial scale (from station # 1, the spring, to station # 6, the closest one to the estuary) and on a temporal scale (the different seasons when samplings were performed) even on a natural ecosystem there may be notable fluctuations that have to be considered while evaluating the sample and the index values. In Figures 3-5 we show the variations of the Quality Classes obtained by the different Indices on the eight rivers examined (River 1 = Albegna; 2 = Arrone; 3 = Tiber; 4 = Volturno; 5 = Mesima; 6 = Ofanto, 7 = Tirso;8 = Simeto).

Among the eight rivers examined, the best water quality is found undoubtedly in the data from the Albegna river, while the Volturno river shows in the study period the worst water quality, compared to the other rivers, and it shows also the widest interval of variation.

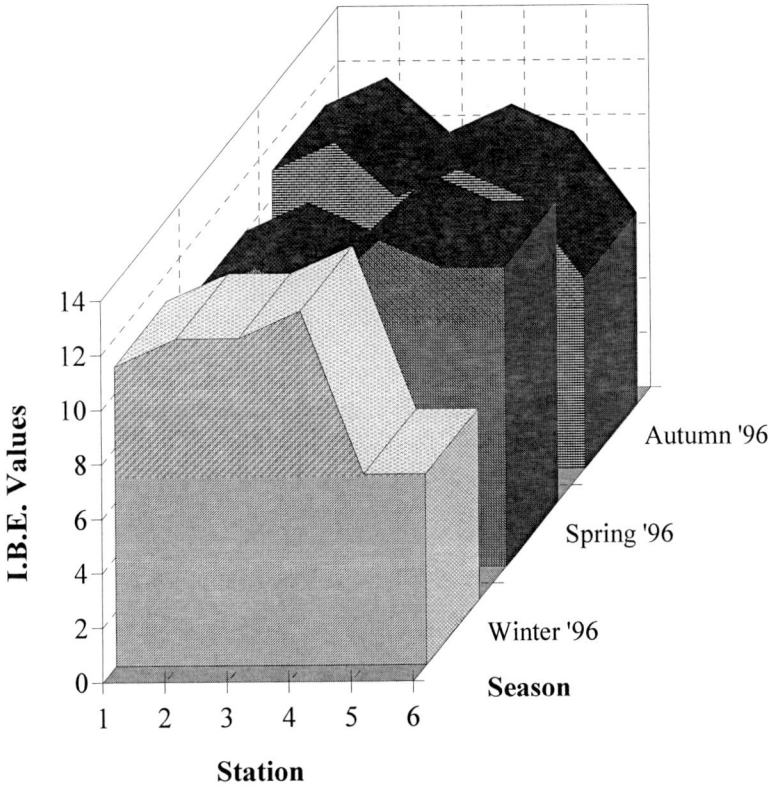

Figure 2. The Albegna River: plot of the IBE values recorded.

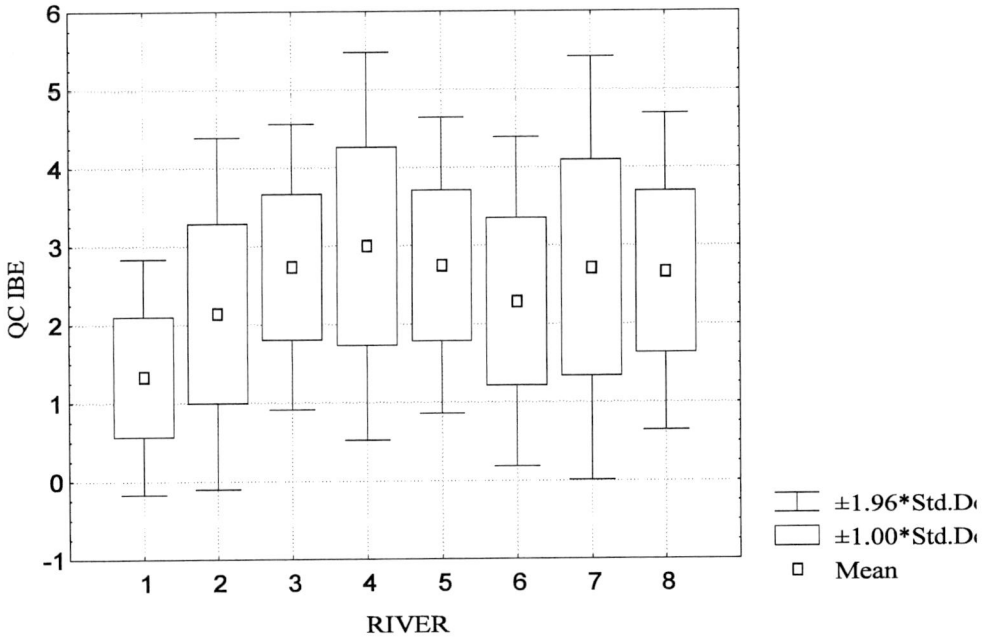

Figure 3. Boxplot of the examined rivers. Variables: QC IBE

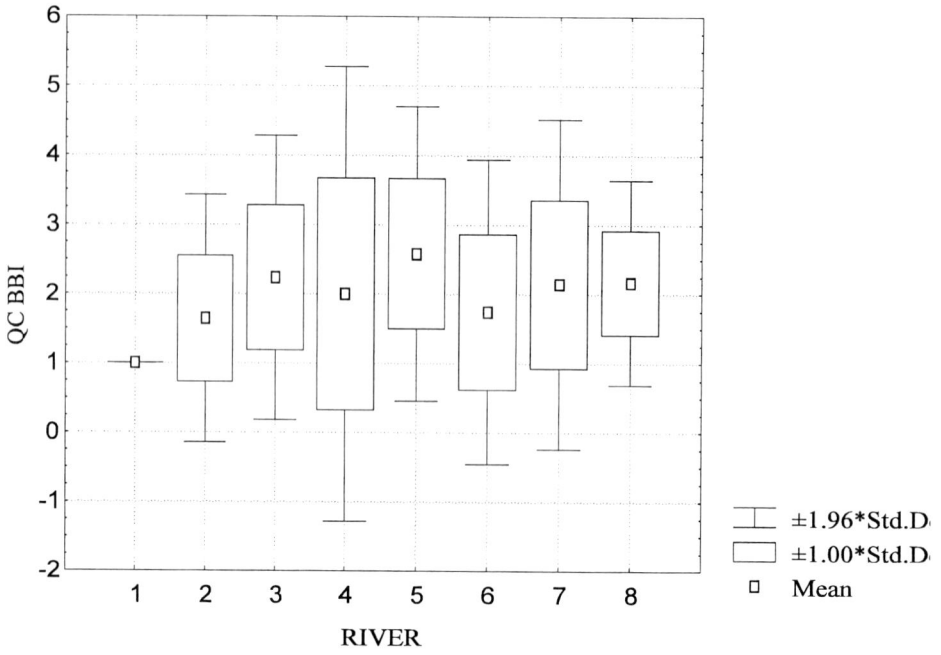

Figure 4. Boxplot of the examined rivers. Variables: QC BBI

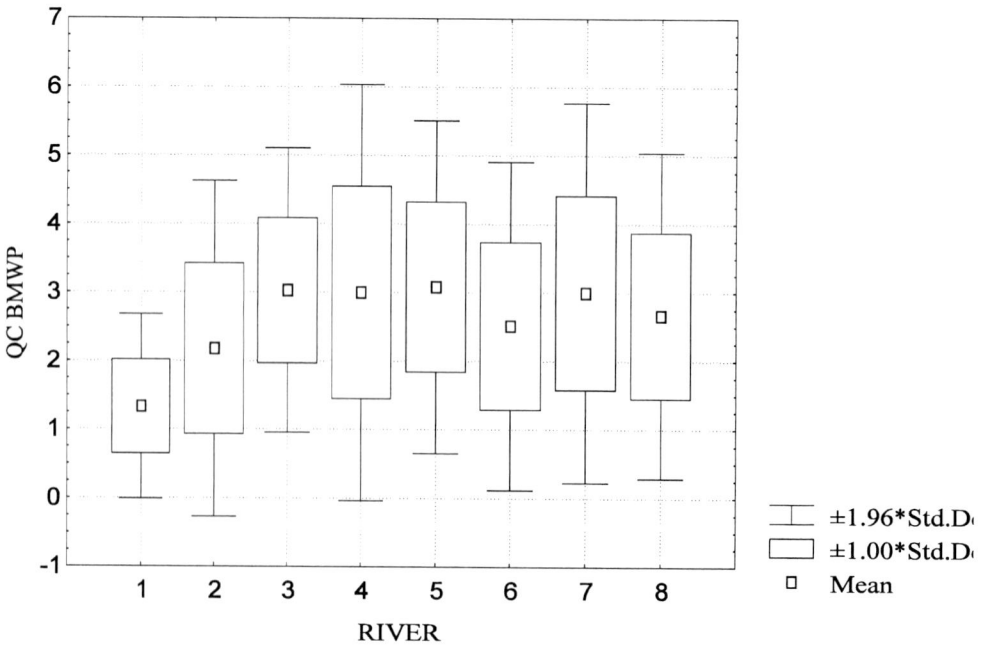

Figure 5. Boxplot of the examined rivers. Variables: QC BMWP.

As a general rule, it can be said that, on the examined Italian rivers, the IBE keeps an intermediate position between the values of the other two indices. The BMWP differs (in Quality Classes) only slightly from the results of the IBE, so much so that, if the Quality Classes are considered, they can be overlapped with a minimum difference. This is confirmed by the Euclydian distances reckoned by means of the Rescaled Squared Euclidean distances.

Single Link
Squared Euclidian Distances

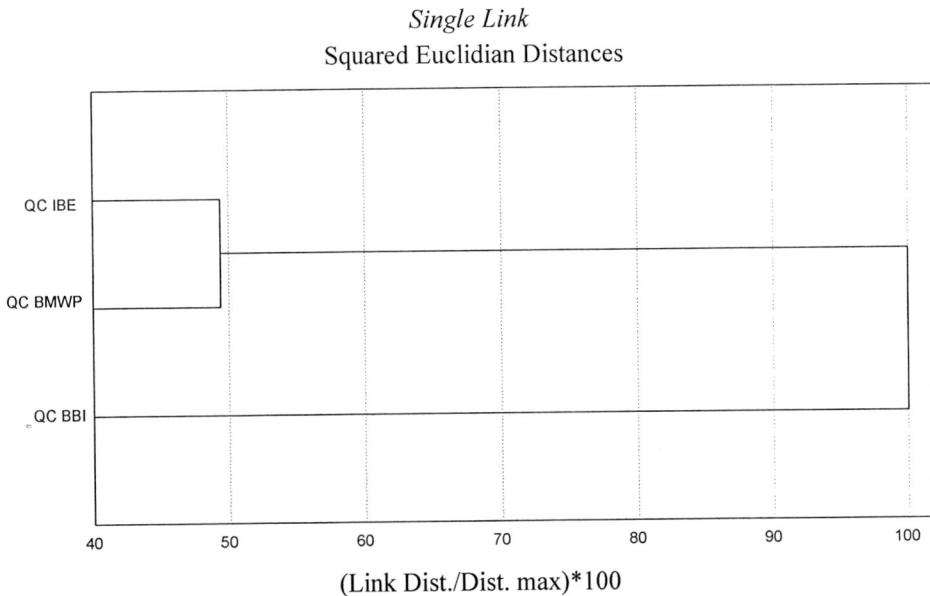

(Link Dist./Dist. max)*100

Figure 6. Three Variables Tree. Quality Classes for the three Indices.

If we consider the single rivers, on the Albegna data set we noted that the BBI was unable to detect small pollution levels, compared to the results of IBE, while the BMWP is enough sensitive to detect even a slight microbiological contamination at the river source, during the spring 1996 (Jacomini *et al.*, 1997, 1998).

On the Arrone River, the BBI still confirms its trend to overestimate Italian rivers quality, while the BMWP shows only light differences with the results of IBE.

Such a trend is confirmed also by the complex situation of a big river like the Tiber. The best stations are confirmed by BMWP, while BBI still retains a trend to understimate the effects of pollution.

In the case of Volturno River, the differences between IBE and BBI are more marked, with only one station confirming the IBE value. The BMWP is less sensitive to the pollution levels, dividing the river in two big classes, one slightly polluted and the other very heavily polluted.

For the Ofanto River, besides an expected overestimation of BBI's Quality Classes, the BMWP oscillates only slightly from the IBE values, without a marked over- or underestimation of the Quality Classes.

On the Mesima River, IBE and BMWP give very similar results. Nothing particular (apart from what said above) can be said for BBI.

The same trends are experienced on the last two rivers, Tirso and Simeto, so that everything that was said before is valid also for these peculiar systems.

3. DISCUSSION AND CONCLUSIONS

The present study is part of the increasing efforts which has been dedicated in the last years to develop a more effective use of macroinvertebrates as monitoring and assessment tools for management of running waters (Buffagni et al., 2000; Lorenz et al, 2004; Czerniawska-Kusza, 2005). The analysed data in this study confirmed the indications of the descriptive statistical methods: if the Quality Classes of the indices are concerned, the IBE and the BMWP are quite close (although the values of the indices are obtained in completely different ways), while the BBI results more distant.

As a general rule, the setting of IBE seems quite good also for the river typologies of Central-Southern Italy. Further refinements should include the seasonal modifications of water quality, due to the lack of controls on environmental protection, and the peculiar situation of some stream.

The authors of this index are already working on this, and probably in a near future there could be an updated version of the index, valid with full marks for the whole Country.

In an European context, the tendency to obtain a standardised methodology for biological monitoring of river quality should follow the guidelines either of IBE, or of BMWP, both consistent with the actual situation of lotic ecosystems. Instead, there has to be noticed the strong bias of BBI, which was set for rivers of different typology and probably with lesser water quality. This could be due to the geographical situation, to the different land use and river quality of Northern Europe, as compared to the situation of Southern streams, and to the lack of further developments and refinements of BBI, as compared to the newest versions of IBE and BMWP.

New data are going to be used to refine this indices analysis, and in a future work this will be accomplished, as said above, utilising also field data and abiotic parameters, allowing the RIVPACS methodology to be compared, too.

ACKNOWLEDGMENTS

This work received a significant contribution from Antonio Pugliese. We would like to thank also Dr. Stefania Capodici for the statistical software collaboration and Dr. Alessandro Massolo for his helpful insights on the statistical elaboration of the data.

REFERENCES

Alba-Tercedor, J. & Sánchez-Ortega, A. 1988. Un metodo rapido y simple para evaluar le calidad biologicas de las aguas corrientes basado en el de Hellawell (1978). *Limnetica 4:* 51-56.

Argano R., Jacomini C., Fornasier F., Sirgiovanni G., Volterra L. & Mancini L., 1996. Aspetti biologici e ipotesi di gestione dei bacini idrografici dei Fiumi Mesima e Petrace (Calabria). *Proceedings of the Conference: "I biologi e l'ambiente oltre il 2000", Venice, 22-23 November, 1996,* Abstract: 17.

Armitage, P.D., Moss, D., Wright, J.F., Furse, M.T. 1983. The performance of a new biological water quality score system based on macroinvertebrates over a wide range of unpolluted running-water sites. *Water Research*, 17: 333-347.

Buffagni, A., Crosa, G., Harper, D. M., & Kemp, J. L. (2000). Using macroinvertebrate species assemblages to identify river channel habitat units: an application of the functional habitat concept to a large, unpolluted Italian river (River Ticino, Northern Italy). *Hydrobiologia*, 435, 213–225.

CEC 2000. Council of European Communities Directive 2000/60/EEC of 23 October 2000 establishing a framework for community action in the field of water policy. Official *Journal of European Communities*, L327/1.

Czerniawska-Kusza, I. (2005) Comparing modified biological monitoring working party score system and several biological indices based on macroinvertebrates for water-quality assessment. *Limnologica* 35 (3): 169-176

Dal Cero C., Mancini L., Di Carlo M., Fornasier F., Bartoni C., Volterra L., 1993. La qualità del Fiume Tevere attraverso analisi multiparametriche. *Ambiente Risorse e Salute*, 19 (XII-VII): 33-41.

De Pauw, N. 1988. Biological assessment of surface water quality: the Belgian Experience. Pp. 197-233. In: La qualità delle acque superficiali, criteri per una metodologia omogenea di valutazione. Provincia Autonoma di Trento, *Assessorato all'Ambiente, Dipartimento Ecologico.*

De Pauw, N. & Vanhooren, G. 1983. Method for biological quality assessment of watercourses in Belgium. *Hydrobiologia,* 100: 153-168.

Ghetti, P.F. 1986. I macroinvertebrati nell'analisi di qualità dei corsi d'acqua. *Manuale di applicazione Indice Biotico: E.B.I. modificato.* Provincia Autonoma di Trento.

Ghetti, P.F. 1995. Indice Biotico Esteso (IBE). *Notiziario dei Metodi Analitici, IRSA (CNR),* ISSN: 0333392-1425: 1-24.

Ghetti, P.F. 1997. Manuale di applicazione. INDICE BIOTICO ESTESO (IBE). I macroinvertebrati nel controllo della qualità degli ambienti di acque correnti. *Provincia Autonoma di Trento, Agenzia Provinciale per la Protezione dell'Ambiente*, Trento.

Ghetti, P.F. & Bonazzi, G. 1980. 3rd Technical Seminar. Biological Water Assessment Methods. Parma, October 1978. *Final Report. Vol. 2. Commission of the European Communities.*

Hellawell, J.M. 1978. Biological Surveillance of rivers. *Water Research Centre, Stevenage,* England.

Jacomini C., Mancini, L., Volterra, L. & Tancioni, L. 1997. Primi dati sulla caratterizzazione ecologica del Fiume Albegna (Toscana, Italia). *Proceedings S.It.E. (Italian Society of Ecology) Congress*, 18: 471-472.

Jacomini, C., Tancioni, L., & Mancini, L. 1998. Ecological Quality of Lotic Ecosystems: the Case-Study of River Albegna. *Proceedings of 7th INTECOL*, Florence, July 1998.

Lorenz, A., Hering, D., Feld, C. K., & Rolauffs, P. (2004). A new method for assessing the impact of hydromorphological degradation on the macroinvertebrate fauna of five German stream types. *Hydrobiologia*, 516, 107–127.

Mancini, 2005. Organization of Biological Monitoring in the EU. In: *Biological Monitoring of Rivers.* Ziglio G., Siligardi M. & Flaim G. (eds). John Wiley & Sons, Ltd, London, 469p.

Mancini, L., Conte, G., Venturi, L. & Tancioni, L. 1998. Le condizioni del Tevere. Pp. 135-142. In: Istituto Ambiente Italia (Ed.), *AMBIENTE ITALIA 1998*, Rapporto sullo stato del Paese e analisi del ciclo delle acque. Edizioni Ambiente.

Mancini, L., Fochetti, R., Venturi, L. & Volterra, L. 1997. Il sistema ambientale del bacino idrografico del fiume Ofanto: aspetti chimico-fisici, chimici e biologici delle acque ed ipotesi di gestione del territorio. *Proceedings of the Conference "Territorio e Società nelle Aree Meridionali", Bari,* 24-27 Oct. 1996.

Mancini L., Fochetti R., Venturi L., Volterra L., 1996a. Il sistema ambientale del bacino idrografico del Fiume Ofanto: aspetti chimico-fisici, chimici e biologici delle acque ed ipotesi di gestione del territorio. *Proc. of Conference "Territorio e Società nelle Aree Meridionali", October 24-25, 1996,* Bari - October 26-27, 1996, Matera.

Mancini L., Jacomini C., Fornasier F., Sirgiovanni G., Belfiore C., Argano R., Fochetti R., 1996b. I macroinvertebrati acquatici dei Fiumi Mesima e Petrace (Calabria). *Proceedings of the Conference: "I biologi e l'ambiente oltre il 2000"*, Venice, 22-23 November, 1996, Abstract: 16.

Mancini M., Dal Cero C., Di Carlo M., Venturi L., Volterra L., Fochetti R., 1995a. I risultati di due campagne di rilevamento eseguite sui Fiumi Tirso e Temo (Sardegna). *Acqua Aria*, 8: 823-828.

Mancini L., Fochetti R., Fornasier F., Venturi L., Dal Cero C., Volterra L., 1995b. La qualità biologica delle acque di sedici fiumi del centro-sud d'Italia (Arno, Chienti, Aniene, Ombrone, Pescara, Biferno, Volturno, Sarno, Ofanto, Basento, Neto, Crati, Simeto, Tirso, Temo). *Biologi Italiani*, XXV (3/95): 25-42.

Mancini L., Dal Cero, C., Gucci, P.B.M., Venturi, L., Di Carlo, M., Volterra, L. 1994. Il Fiume Volturno - La definizione dello stato di salute del corpo idrico. *Inquinamento,* 36 (10): 70-83.

Mancini L., Bernabei S., Iavarone I., Volterra L., Di Girolamo I., 1993. La qualità ambientale del fiume Arrone. *Ambiente Risorse e Salute XII/9*(21):34-37.

Tittizer, T.G. 1976. Comparative study of biological-ecological water assessment methods. Practical demonstration of the River Main (2-6 June, 1975). *Summary Report. Comparisons of biological-ecological procedures for assessment of water quality.* In Amavis, R., Smeets, I. (Eds.)

Tuffery, G. & Verneaux, J. 1968. Méthode de détermination de la qualité biologique des eaux courantes. Exploitation codifiée des inventaires de la faune du fond. Ministère de l'Agriculture (France), *Centre National d'Etudes techniques et de recherches technologiques pour l'agriculture, les forêts et l'équipment rural "C.E.R.A.F.E.R.", Section Pêche et Pisciculture.*

Woodwiss, F.S. 1964. The biological system of stream classification used by the Trent River Board. *Chemistry and Industry*, 14: 443-447.

Woodwiss, F.S. 1978. Comparative study of biological-ecological water quality assessment methods. Second practical demonstration. Summary Report. Commission of the European Communities.

Woodwiss, F.S. 1980. *Biological monitoring of surface water quality. Summary Report. Commission of the European Communities.* ENV/787/80-EN.

Wright, J.F., Armitage, P.D., Furse, M.T., Moss, D. 1989. Prediction of invertebrate communities using stream measurements. *Regulated Rivers: Research and Management,* 4: 147-155.

Wright, J.F., Moss, D., Armitage, P.D., Furse, M.T. 1984. A preliminary classification of running water sites in Great Britain based on macro-invertebrate species and the prediction of community type using environmental data. *Freshwater Biology,* 14: 221-256.

INDEX

B

C

D

Q

T

Z

DATE